EBEN SMITH

Mining the American West

SERIES EDITORS: DUANE A. SMITH | ROBERT A. TRENNERT | LIPING ZHU

Boomtown Blues: Colorado Oil Shale | ANDREW GULLIFORD

Eben Smith: The Dean of Western Mining | DAVID FORSYTH

From Redstone to Ludlow: John Cleveland Osgood's Struggle against the United Mine Workers of America | F. DARRELL MUNSELL

Gambling on Ore: The Nature of Metal Mining in the United States, 1860–1910 | KENT A. CURTIS

Hard as the Rock Itself: Place and Identity in the American Mining Town | DAVID ROBERTSON

High Altitude Energy: A History of Fossil Fuels in Colorado | LEE SCAMEHORN

A History of Gold Dredging in Idaho | CLARK C. SPENCE

Industrializing the Rockies: Growth, Competition, and Turmoil in the Coalfields of Colorado and Wyoming | DAVID A. WOLFF

The Mechanics of Optimism: Mining Companies, Technology, and the Hot Spring Gold Rush, Montana Territory, 1864–1868 | JEFFREY J. SAFFORD

Mercury and the Making of California: Mining, Landscape, and Race, 1840–1890 | ANDREW SCOTT JOHNSTON

The Once and Future Silver Queen of the Rockies: Georgetown, Colorado, and the Fight for Survival into the Twentieth Century | CHRISTINE A. BRADLEY AND DUANE A. SMITH

The Rise of the Silver Queen: Georgetown, Colorado, 1859–1896 | LISTON E. LEYENDECKER, DUANE A. SMITH, AND CHRISTINE A. BRADLEY

Santa Rita del Cobre: A Copper Mining Community in New Mexico | CHRISTOPHER J. HUGGARD AND TERRENCE M. HUMBLE

Silver Saga: The Story of Caribou, Colorado, Revised Edition | DUANE A. SMITH

Thomas F. Walsh: Progressive Businessman and Colorado Mining Tycoon | JOHN STEWART

Yellowcake Towns: Uranium Mining Communities in the American West | MICHAEL A. AMUNDSON

EBEN SMITH

The Dean of Western Mining

DAVID FORSYTH

UNIVERSITY PRESS OF COLORADO
Louisville

Published by University Press of Colorado
245 Century Circle, Suite 202
Louisville, Colorado 80027

Manufactured in the United States of America

ASSOCIATION of UNIVERSITY PRESSES The University Press of Colorado is a proud member of the Association of University Presses.

The University Press of Colorado is a cooperative publishing enterprise supported, in part, by Adams State University, Colorado State University, Fort Lewis College, Metropolitan State University of Denver, Regis University, University of Colorado, University of Northern Colorado, University of Wyoming, Utah State University, and Western Colorado University.

∞ This paper meets the requirements of the ANSI/NISO Z39.48–1992 (Permanence of Paper).

ISBN: 978-1-64642-178-7 (hardcover)
ISBN: 978-1-64642-179-4 (ebook)
https://doi.org/10.5876/9781646421794

Library of Congress Cataloging-in-Publication Data

Names: Forsyth, David, 1977– author.
Title: Eben Smith : the dean of Western mining / by David Forsyth.
Other titles: Mining the American West.
Description: Louisville : University Press of Colorado, [2021] | Series: Mining the American West series | Includes bibliographical references and index.
Identifiers: LCCN 2021021144 (print) | LCCN 2021021145 (ebook) | ISBN 9781646421787 (hardcover) | ISBN 9781646421794 (ebook)
Subjects: LCSH: Smith, Eben (Ebenezer), 1831–1906. | Miners—Colorado—Cripple Creek—Biography. | Mining engineers—Colorado—Cripple Creek—Biography. | Gold mines and mining—Colorado—Cripple Creek—Biography. | Silver mines and mining—Colorado—Cripple Creek—Biography. | Mineral industries—Colorado—Cripple Creek—Biography. | Cripple Creek (Colo.)—History—19th century.
Classification: LCC TN140.S59 F67 2021 (print) | LCC TN140.S59 (ebook) | DDC 622.092 [B]—dc23
LC record available at https://lccn.loc.gov/2021021144
LC ebook record available at https://lccn.loc.gov/2021021145

Cover photograph: Eben Smith, ca. 1875 (History Colorado, Object # 89.451.4395)

For the Fantastic Forsyths

Contents

Acknowledgments

In 1990, the late L. Douglas Hoyt, then-owner of the Smith Mansion at Eighteenth and York in Denver, gave my family and me a tour of the house. It was a wonderful chance to see the inside of the house that started my interest in Eben Smith, and I have always appreciated his willingness to indulge a kid interested in history.

This book began life as my master's thesis at the University of Colorado at Denver many years ago, and I wish to thank James Whiteside, Rebecca Hunt, Jay Fell, and Tom Noel for their advice and suggestions as I worked on it. I am also grateful to Duane Smith at Fort Lewis College for his suggestions during my research.

A number of other people were very helpful in my research. Jim Prochaska and Alan Granruth at the Gilpin Historical Society dug out pictures and papers for both Eben Smith and his granddaughter Emmy Wilson, who owned the Glory Hole Tavern in Central City. The staff at the Denver Public Library's Western History Department brought out box after box of Eben Smith's papers along with Moffat Estate Company papers and helped point me in the direction of more sources. The staffs at the Carnegie Branch Library in Boulder, the Lake County Public Library in Leadville, and the Stephen H. Hart Library at History Colorado in Denver were also of great help to me. Librarians throughout the country helped me find newspaper articles and obituaries as I tracked the Smith family. I would like to particularly thank D. Cohen at the Sacramento Public Library and the staff at Mountain View Cemetery in Oakland, California, whose help allowed me to discover the first name of Eben Smith's first wife, Caroline, who had always been referred to as Mrs. Jordan or Mrs. Smith in sources.

At the University Press of Colorado, I would like to thank the anonymous peer readers, Jessica d'Arbonne, Charlotte Steinhardt, Rachael Levay, Beth Svinarich, Laura Furney, Dan Pratt, and Darrin Pratt for their suggestions, help, and guidance in getting this book published. Sue Carter helped me edit the manuscript when I thought I could not adjust one more comma or change one more word after so many years of working on it, and Cheryl Carnahan helped polish it into the final version.

I also owe a very special thanks to Eben Smith's family. His great-granddaughters Betty Works and Barbara Landweer Smith and great-great-granddaughters Nancy Smith Piano, Leslie Wheary, Julie Mynett, and Michel Tritt provided me with a vast amount of family history, stories, and photographs. Anne McCarroll and her father, Louis O'Kane, both of Ireland, whose grandmother and mother, Fanny Watson O'Kane, was good friends with Cora Smith Carnahan, were also wonderful sources and provided me with copies of many letters and photographs.

Finally, I am thankful to my family for their constant support and for putting up with Eben Smith for so many years.

EBEN SMITH

Introduction

In July 2012, more than 100 years after Eben Smith's death, the owners of a warehouse that once belonged to the Denver Mine and Smelter Supply Company began to demolish the building with the hope of selling the land to a new owner who wanted to construct apartment buildings. Neighbors of the warehouse, which was built in 1909, filed an ultimately unsuccessful petition to have the warehouse declared a Denver landmark and be saved from destruction. The neighbors argued that the building deserved landmark status because of its connection to Colorado's mining history and to mining pioneer Eben Smith, overlooking the fact that Smith had sold his interest in the Denver Mine and Smelter Supply Company in 1901 and died in 1906, three years before the warehouse was built. Critics vocally condemned the landmark attempt, arguing that allowing a hostile neighbor to attempt to thwart an owner's plans for a historic building made a mockery of Denver's landmark preservation ordinance, and *Denver Post* columnist Vincent Carroll pointed out the folly of using Smith's name to save a building he had never seen.[1]

DOI: 10.5876/9781646421794.c000

My interest in Eben Smith started with a house. I had passed it while on a school field trip in 1990, and it fascinated me. Since I could only remember the area where I had seen the house, my dad, brother, and I spent a Sunday afternoon searching for it. When we finally found it, I learned from a sign on the fence that it was called the Smith Mansion. Over the following months and years, I visited the Denver Public Library a number of times to find out whatever I could about the house and its history. I even had the opportunity to take a tour of the house, conducted by then-owner L. Douglas Hoyt. As part of my research on the history of the house, I also came to know about the family of its builder and first owner, Eben Smith's son Frank L. Smith, who had made his own fortune from mining and built the house in 1902.

While the story of Frank Smith and his house was compelling, even more impressive was the story of his father, Eben Smith. Like many in 1849 and 1850, Smith went to California during the gold rush to seek his fortune. Unlike the majority, he found it. He next turned his sights on Colorado, arriving there in 1860 with a new business partner, Jerome Chaffee. Over the next forty-six years, Smith was involved in mining in nearly all of the major mining camps in the state, from Central City to Cripple Creek, and he developed mines such as the Bobtail, Little Jonny, and Victor. One short article on his life said "the name of the fabulously rich mines which he discovered or developed . . . read like chapter headings in the romance of the West." In addition to Chaffee, Smith was also associated with David Moffat, earning a reputation as his "right-hand" man. The mining teams of Smith and Chaffee and Smith and Moffat were among the most successful and respected in Colorado, and many in the state held Smith in especially high regard.[2]

Just as the names of his mines read like chapter titles in the history of the West, so do the names of Smith's associates. In addition to Chaffee and Moffat, just a few of his better-known associates were Horace Tabor, J. J. Brown, Bela Buell, and Charles J. Hughes. Smith met John Gregory and William Green Russell, who would become legendary figures in Colorado mining, shortly after they found gold in Colorado. And while not as well-known, many other Smith associates and employees would appear in mining camps around the state, including Charles Mater, several members of the Curnow family, and dozens of other miners.

Without mining, the state of Colorado would not exist as it does today, and Eben Smith played a major part in developing mining in Colorado. Yet despite the credit for establishing Colorado's mining industry that he received during his lifetime, Smith has received little attention from historians. In part, this reflects the kind of life he lived. Smith was more than happy to leave the role of public promoter and the responsibility of being in the public eye to his partners while he worried about the output of his mines. For the mines to give up their riches, they needed people to invest money. Investors such as Horace Tabor, J. J. Hagerman, David Moffat, and Jerome Chaffee supplied the capital, but they needed men like Eben Smith to coax the mines into producing. The mining and milling knowledge he acquired in California and his ability to put it into practice in Colorado made him a leader in technological innovation, and his willingness to use new technology usually served him well. Smith indeed could make the mines produce.

Steven F. Mehls, in his biography of David Moffat, writes that Moffat "took the more prestigious but in some ways rather risky route of the mining financier and promoter." The same was true of Leland Stanford, who invested in mines but left the day-to-day operations to others. Horace Tabor first made his fortune by grubstaking the discoverers of the Little Pittsburgh Mine in Leadville. J. J. Hagerman also made his fortune investing in mines in Michigan and then in Cripple Creek. The Guggenheims, already rich from Meyer Guggenheim's numerous business ventures, built their great fortune, according to biographer John H. Davis, from investing in mines in Leadville and Cripple Creek among other places, as well as in smelters in Colorado and Mexico.[3]

For these men, mining was a means to an end. In each of these cases, they used the fortunes they made in mining to pursue their other and, to them, more important interests. In Moffat's case, this was to build railroads and promote Colorado. Stanford and Hagerman also used their mining riches to build railroads. Tabor put his money to more artistic pursuits, building his famed opera houses in Leadville and Denver, in addition to the office building known as the Tabor Block in Denver. The Guggenheims used their fortune to establish a number of charitable foundations. Chaffee used his fortune to advance his political career and promote the state.[4]

Mentioned far less often in biographies are those men who actually worked the mines. Some made it; some did not. One who did not was Arthur Hill, who went to Alaska in 1898 and to Arizona in 1899 to pan for gold. His biographer, John Moring, described Hill as "notional," given the many interests he had in life. Hill later found success as an actor and a police officer in San Diego, not as a miner. Another example of a man who actually worked the mines was Winfield Scott Stratton. Arriving in Colorado in the late 1860s, Stratton spent seventeen years prospecting all over the state, always coming up empty-handed. Stratton became so obsessed with finding a mine that he gave up a promising career as a carpenter. Unlike Hill, Stratton finally achieved his dream when he discovered the famous Independence Mine in Cripple Creek in 1891 and became a millionaire. Thomas F. Walsh, another Smith contemporary, prospected for years in the Black Hills and Colorado before discovering the Camp Bird Mine near Ouray, which made him a millionaire.[5]

Eben Smith straddled the positions of investor and worker. Starting out at the bottom ranks with Hill and Stratton, Smith soon worked his way up to the ranks of Moffat and Tabor. Smith's life and career might best be compared to two of his fellow Central City pioneers, Henry Teller and Peter McFarlane. Arriving in Central City in 1861, Teller started out as a small-time lawyer who worked his way up to become a US senator and secretary of the interior under Chester A. Arthur. By the time of his death, Teller was one of the most respected men in Colorado. Eben Smith started out much the same way, although he arrived in Central City already well-off. Just as Teller rose through the ranks of Colorado politics, Smith rose through the ranks of mining. But he never left behind his days of digging in the dirt. For Eben Smith, mining was his great love.[6]

Peter McFarlane arrived in Central City in 1869, at the tail end of a depression in mining brought about by the difficulty in processing ore from the area. McFarlane and his brothers worked in construction before moving into designing and building mining equipment. Along the way, he served as a member of the city council in Central City and as the town's mayor. By the late 1890s, McFarlane's long experience had given him a strong reputation in the mining community and the state as a whole. When Gilpin County's mining industry began to decline in the early 1900s, McFarlane remained convinced that it could be revived with the right equipment. As the owner of the Central City Opera House from the late 1890s until his death in 1929, McFarlane clung to the building even though he lost a great deal of money in doing so. Given the troubles Smith had with some of his business ventures near the end of his own life, he might very well have sympathized with McFarlane's troubles.[7]

Much of Eben Smith's career intersected with America's Gilded Age. In the years after the Civil War, the country found itself rapidly changing from an agrarian economy to one based on industry. As a result, people began to move from farms and rural areas to the big cities. J. P. Morgan, Andrew Carnegie, and John D. Rockefeller, among others, earned fabulous sums of money that they used to build elaborate homes and fund charitable institutions, all while becoming either heroes or demons to the world around them. In the frontier West, natural resources such as gold, silver, and land became hot commodities ripe for exploration and exploitation as tens of thousands of people headed there to mine, speculate in land, build railroads, or supply those who were heading there.[8]

Historian Gene M. Gressley describes westerners of the Gilded Age as possessing a "powerful opportunistic sense . . . If economic rewards did not come today, they would tomorrow or the morrow after that." No matter what happened, westerners were convinced that a bright future awaited them despite any setbacks or failures they encountered. While westerners are often viewed as fiercely independent, Gressley writes that the West matured only after absorbing "massive amounts of Eastern capital and technology." When westerners exceeded their economic reach, they relied on the federal government to intervene in areas such as transportation and water. In the West of the late nineteenth century, the public and private sectors, according to Gressley, were expected to cooperate. Westerners also imported eastern legal, political, and social institutions as they created new societies based on what they had known. But while seeking help from the East, westerners also feared being dominated by easterners. Gressley argues that the relationship between the West and the East during the Gilded Age was much like that of a colony and a mother country.[9]

While Gressley's colonial argument has fallen out of favor in recent times, the optimism displayed by the miners has not. Historian Duane A. Smith writes that "optimism was never absent" in Colorado's mining industry. Eben Smith's life illustrates the opportunistic sense and optimism Gressley and Duane Smith write about. That search for a brighter future and for more economic blessings than his parents had drove Eben Smith throughout his career. In all, about 250,000 people arrived in California after James Marshall, an employee of John Sutter, discovered gold near Sacramento on January 24, 1848. This is the story of Eben Smith, one of those 250,000, who achieved the success of which many of them only dreamed.[10]

1

"I Left Pennsylvania When I Was 19"

In August 1902, entrepreneur, miner, and politician Eben Smith wrote what amounted to his only autobiography. To prove to a man in Michigan that he had never been to Michigan and therefore could not possibly owe him thirty-two dollars, he wrote, "I left Pennsylvania when I was 19 years old, went to California, remained there until 1859, returned to Pennsylvania, came to Colorado in 1860 and have lived here ever since." While this short life history hit the high points and saved Smith thirty-two dollars, it left out a great deal.[1]

According to a family history compiled between 1889 and 1891 and published in 1892, the Smith family was of "sturdy Scotch-Irish ancestry." Eben Smith's great-grandparents, John and Mary Smith, lived in Ireland. A family friend interviewed for the family history said, "The Smiths were a well to do people in the old country" based on the high position of their church pew. Eben Smith's grandparents, Samuel and Jane Smith, along with Samuel's brothers William, Robert, and James, left Ireland for the United States in 1796. Each of the four brothers bought a farm in Erie County, Pennsylvania, with Samuel and Jane settling on 400 acres near Union Township. Samuel and Jane

DOI: 10.5876/9781646421794.c001

Smith had twelve children, the third of which was their son William, born in 1801. In 1817 the Smith family moved to a 250-acre farm near Beaverdam. William Smith married Mary Nelson in Beaverdam on February 15, 1831.[2]

Like his father, William Smith was a farmer. The oldest of William and Mary's eight children, Ebenezer Smith, was born in Beaverdam, Pennsylvania, on December 17, 1831. According to a later biography, Smith inherited "little from his parents save a vigorous physique and strong mental powers." A reporter for the *Erie Daily Times* wrote in 1906 that Smith also had a "determined personality" that would ensure success. Little is known about Smith's early life, but he was educated in public schools in Erie and at a private academy in Waterford, Pennsylvania. He also trained as a ship's carpenter, which some thought helped him greatly in his future mining career.[3]

After James Marshall discovered gold at Sutter's Mill in California in 1848, it seemed as if the entire population of California, from Los Angeles to San Francisco, abandoned their cities to go search for gold. As word of the discovery spread farther, people from the United States and the world began heading to California. By land and by sea, thousands made their way to California seeking wealth and adventure. H. W. Brands writes that for many people, "the news from California was the most exciting" they had ever heard, and the "rush to California promised to be the event of their lifetime." For many of those who headed there, the trip to California was the first long trip of their lives, and no matter how they chose to go it was indeed an adventure.[4]

Nineteen-year-old Eben Smith was not immune to the allure of the discovery of gold in California, and in 1850, he decided to go there to try his hand at mining. He traveled "by vessel by way of the isthmus" and arrived in San Francisco in 1851. The trip Smith undertook required taking a ship from New York to Panama, crossing the isthmus, and then taking a second ship from Panama to San Francisco. The voyage cost between $200 and $500, meaning that Smith must have been able to save a fair amount of money in order to take it. Gold seekers first took a ship to Chagres, the port on the Caribbean side of Panama. The mass of humanity passing through on their way to California overwhelmed the town, with people sleeping wherever they could, which increased the risk of catching malaria or other diseases. In addition, criminals preyed on the travelers. The gold seekers then had to cross the isthmus itself.[5]

From Chagres, gold seekers traveled on the Chagres River to Gorgona by steamship or other watercraft. They made the final trek from Gorgona to Panama City by horse, by mule, or on foot. The path was little more than a dirt track that frequent rains turned to mud, making the journey even more difficult. In late 1849, those heading to Panama City met those heading back to Chagres, done with California and its promise of riches. Once in Panama City, the early fortune seekers faced a shortage of ships because shipping companies were overwhelmed by the number of people heading to California. The wait for a spot on a ship could sometimes stretch into weeks, leading some to wonder if the journey was worth the trouble. In all, the journey usually took about two to three months (for those who could not afford to travel through Panama, it took four to eight months to travel around Cape Horn).[6]

After leaving San Francisco, Eben Smith first settled at Grass Valley in Nevada County, most likely engaging in placer mining. He then moved on to Sierra County, where he panned for gold for a short time before returning to Nevada County. His placer mining ventures must have been somewhat successful, as his next venture was building a hotel at French Corral in Nevada County. The *Sacrament Daily Union* reported in 1852 that French Corral was thriving "in a degree unprecedented," so much so that when a fire destroyed a large part of the town in 1853, it caused over $40,000 in damage. By then Smith had sold his hotel, however, and had moved on to Placer County, where he took a job at the Texas Quartz Mill, which was located near the town of Damascus. The mill was reportedly the largest of its kind in California, and Smith's stepson Fletcher Jordan said his training as a ship's carpenter gave him a "natural aptitude for all work about a mill." He appeared to be an "ordinary sort of fellow," but his "natural ability rendered him useful and popular and he got on well." An efficient amalgamator, Smith became a master of the mineral-treating process and soon bought a half interest in the mill. He continued to work at the mill in addition to mining in various parts of Placer County for about a year.[7]

Smith may have first encountered quartz mills while working in Grass Valley, where residents spent much of the 1850s developing the quartz mining and milling industries. As far as most miners were concerned, quartz mining was steadier than placer mining, but one problem always remained: separating the gold from the quartz in which it was found. According to historian Ralph Mann, miners were convinced that whoever could solve the problem of separating the gold from the quartz would become enormously wealthy, and the challenge attracted "skilled mechanics and draftsmen" to Grass Valley to give it a try. No matter who was trying to solve the problem, though, Mann writes that "the basic combination necessary for a successful quartz mine remained constant: a rich claim, skilled management, a high-quality mill, capital adequate to overcome unforeseen accidents and delays, and good luck."[8]

Quartz, or stamp, mills crushed ore under iron blocks (stamps) that were attached to wood or iron rods connected to a rotating beam at the top of the mill. The stamps rose and fell based on the rotation of the beam, crushing the ore into a fine powder as it passed under them. A stream of water then washed the crushed ore over copper plates that were impregnated with mercury. The mercury and gold formed an amalgam that was then cooked in a retort, burning off the mercury and leaving the gold behind. Historian James E. Fell Jr. writes that the "principle was simple, but stamp milling required a knowledge of engineering, a supply of semiskilled labor, and access to capital." The arrival of stamp mills in California and, later, in Colorado was an important moment for the mining industry in both places.[9]

Learning by experience and watching others was not uncommon among early miners in California. Historian H. W. Brands refers to the way miners learned from each other as "empirical science." Miners watched each other to try to determine what techniques seemed to produce better results and what conditions made one claim pay well while another failed. They also studied rock formations and soon developed the knowledge of which would pay and which would not. In addition, "California revealed what gold geology looked like," enabling miners to search for it around the world. Where existing techniques did not work to remove gold from rock, they had to invent new ones. While quartz milling required more expert knowledge of chemistry, physics, geology, and engineering, for an interested and willing learner, observation was still a good way to learn the process, and Smith took it all in.[10]

While working at the mill, Smith took his meals at a boardinghouse in Damascus owned by Caroline Jordan, a twenty-six-year-old widow and mother of four from Iowa. George Jordan was a farmer in Dubuque, Iowa, who died in May 1850, leaving his wife, Caroline, with four young children.

FIGURE 1.1. Eben Smith, seated in the front row, ca. 1860. The woman seated next to him is possibly his wife, Caroline. Michel Tritt Collection.

Although Caroline remained in Iowa for several months after her husband's death, she eventually decided to move the family to California. According to her son Fletcher, the family took the much more difficult overland route rather than going by ship. They settled in Damascus, where Caroline opened a boardinghouse and met Eben Smith. In late 1852, Smith and Caroline traveled to Georgetown, California, to get married.[11]

Most of the people who went to California during the gold rush never intended to stay there. They planned to make a quick fortune and return to the homes and lives they had left behind, resulting in the often transient nature of early mining camps. For those who did stay, not only did they have to work at getting gold, they also had to work at creating a new society. Gene Gressley writes that most western towns attempted to re-create the eastern society their residents had known. Brick buildings, which could include churches, theaters, hotels, and a post office, were one sign of permanence and stability. Stable town governments, a newspaper, the arrival of professionals such as doctors and lawyers, and growing families were additional signs that a new society was taking hold. Although Eben Smith would play a much larger part in establishing a permanent society in Colorado, his brief stint as a hotel owner and his marriage were small steps he took to help build a new society in California.[12]

Smith also helped establish a stable society in California through his involvement with quartz mills. According to Ralph Mann, those who advocated for a stable society pointed to the quartz mills as a further indicator of that stability. In Grass Valley, observers often said the presence of the quartz mills gave the camp the appearance of a New England factory town and referred to it as a village rather than a camp. The mills, they argued, "created stable employment and therefore a stable society." That regular employment would further "impose order and discipline on the laboring populations and create a permanent management class."[13]

Eben and Caroline Smith and her four children moved to Iowa Hill in 1853 or 1854. By early 1854, Iowa Hill was a major center of mining activity, with the *Sacramento Daily Union* reporting in May of that year that anyone who attempted to mine within 6 miles of the town was "abundantly remunerated." Smith formed a partnership with William McMertie and William Walsh to engage in mining and build a quartz mill. Smith was superintendent of construction and half owner of the mill, and when it was completed it overtook the Texas Quartz Mill as the largest milling plant in California, employing more than 300 men. One year later, Eben Smith and R. A. McClellan bought the entire property in addition to the already prosperous Pioneer Ledge and Mammoth Lodes. Together, they worked the property until May 1859, when Smith sold his interest to McClellan at an "advantage" according to Fletcher Jordan. By then, Eben and Caroline had two sons, twins Samuel and Lemuel, who were born in 1858. Fletcher Jordan said Smith was financially comfortable after selling the mill, and he decided it was time to take a break from California. The family first went to Dubuque, Iowa, and then to St. Joseph, Missouri, to visit Smith's brother Samuel, who was a doctor there. Smith originally planned to go on to Pennsylvania to visit his family and then return to California, but the visit to St. Joseph changed his plans.[14]

While visiting his brother, Eben Smith frequently conducted business in the banking house of Lee and Chaffee. He was "quite a conversationalist," and his favorite subject was mining. In the bank, he struck up a conversation about mining with one of the bank's owners, Jerome Chaffee. Chaffee had been born on his father's farm near Lockport, New York, on April 17, 1825. For the first seventeen years of his life he lived there, working on the farm in the summers and attending public school in the winters. In addition to studying at school, he "supplemented" his knowledge with "hard study at home." When he was seventeen, he moved with his parents to a farm near Adrian, Michigan, where he attended a private academy. Then, at nineteen, he moved to Indiana, where he taught school and clerked in a store.[15]

Two years later, Chaffee returned to Adrian, again clerking in a store before becoming a bookkeeper at a bank. He rose rapidly in the bank, and when it failed, he served as receiver to take care of the bank's unfinished business. In 1848, he met and married Miriam Comstock. In 1857, Chaffee went to Kansas Territory to help found the town of Elwood before moving to St. Joseph later that year to found a bank with Albert Lee. In January 1857, the Chaffees' daughter Fannie was born, the only one of their four children to survive childhood. The following November, Miriam Chaffee died. Left with an infant daughter, Chaffee was looking for a business venture that would be financially rewarding when he met Eben Smith in the bank.[16]

The nationwide depression of 1857 hit the Missouri River Valley especially hard, and reports of gold discoveries in what would become Colorado attracted a great deal of attention. While Smith was in St. Joseph, he met several of the men who had been searching for gold in the Rocky Mountains, including John Gregory, William Green Russell, John Lyons, and Dick Milsap. All were happy to share the stories of what they had found with the experienced miner. Smith and his new friend Chaffee spent a great deal of time discussing the gold find, and he became so intrigued that he canceled his planned trip to Pennsylvania and eventual return to California to join the Colorado gold rush.[17]

Eben Smith's experience at the two quartz mills in California made him an expert on milling ore, and the reports from Colorado convinced him that a stamp mill would be a welcome and profitable addition to the mining activities there. Chaffee was "always awake to the possibilities in a money-making way," and he was excited when Smith told him what they could do with a mill in the region. The two quickly formed a partnership to have a stamp mill built and shipped to Colorado, but whether Chaffee or Smith convinced the other to join forces is unclear. Some sources suggest that Samuel Smith persuaded his brother to take Chaffee on as a partner even though he had no mining or milling experience.[18]

After forming their partnership in early 1860, Smith and Chaffee had a twelve-stamp quartz mill built at Leavenworth, Kansas. Fletcher Jordan believed Chaffee put up most of the money for construction, but he and Eben Smith were equal partners from the beginning. Once the mill was complete, they had it shipped across the plains to Colorado in preparation for their arrival. Chaffee arrived in Colorado in February 1860, followed by Smith and his family (including his brother J. Nelson) on May 26, 1860. As soon as both partners were in Colorado, they surveyed the situation and decided to erect the mill, as well as the building to house it, on Lake Gulch, southeast of the boomtown of Central City in what would become Gilpin County.[19]

2

"One of Colorado's Headest Men"

The Panic of 1857 guaranteed that reports of gold in the Rocky Mountains reached an eager audience. Two parties of prospectors, one led by William Green Russell and the other led by John Beck, met up at the Arkansas River and made their way to the South Platte, where they met a third group from Missouri. In late June and July the three groups panned for gold in Cherry Creek and Ralston Creek, without success. When they failed to find gold, all but 13 (of the original 100) of the men left. But on July 7 or 8, 1858, those remaining 13 found small amounts of gold in Little Dry Creek. News of the discovery slowly spread to towns close to the new goldfield, and hundreds of people (especially those without jobs or businesses) ventured to the new towns of Denver and Auraria to try to find their own fortunes.[1]

Placer deposits, which those early miners worked, run out fairly quickly. For the new gold rush to last, someone needed to find lode gold. Several prospectors went looking for it, including John Gregory, an experienced miner from Georgia who was on his way to British Columbia but decided to see the gold discovery at Little Dry Creek for himself. In January 1859 he started

DOI: 10.5876/9781646421794.c002

prospecting in canyons around the area, finally reaching present-day Black Hawk in April. He was certain he had discovered gold, but a severe snowstorm drove him back to Golden and he was unable to investigate his find for several weeks. Finally, at the beginning of May, he made his way back to his earlier find, and on May 6, he staked two claims on what became known as the Gregory Lode. Between May 16 and May 23, he pulled $972 worth of gold from the claims and then sold them for $21,000. A week after his initial discovery, people in Denver reported seeing a vial holding $80 worth of gold found by Gregory at his diggings.[2]

Within a few weeks of Gregory's discovery, 20,000 to 30,000 people were prospecting in the hills around Gregory Diggings, as the area came to be known. This number quickly dropped to between 4,000 and 5,000 when the miners realized that most of the gold was locked in quartz formations and it would require a great deal of hard work and equipment to remove it. Thousands of would-be miners either headed for home or to other potential sites in the region. The *Rocky Mountain News* reported on August 13, 1859, that there were almost 300 buildings in the town, and 2,800 to 3,000 people were living there. One visitor described the cabins and other buildings as seeming to hang impossibly from the canyon walls. When the first snow fell the following September, the population was cut almost in half as more people abandoned the camp and headed for home.[3]

By the time Eben Smith and his family arrived in the area around Gregory Diggings in 1860, the year before Colorado became a territory, the area had become somewhat less primitive than it was the year before. Several mining camps had grown up around the original camp at Gregory Diggings, including Nevada City (later Nevadaville), Springfield, Dog Town, and Lake Gulch, where Smith and Chaffee installed their stamp mill. Each mining district had its own rules for handling mining claims and dispensing justice. The area that included Gregory Diggings eventually became Mountain City. Central City was uphill, between Mountain City and Nevada City, while the area below Mountain City became Black Hawk. Mountain City was initially the most important of the four cities, but Central City soon overtook it due to its central location, and in 1864 Mountain City became part of Central City.[4]

Smith and his family most likely arrived in Central City on a stagecoach, the first of which started running to the mining camps in early 1860. The route ran from Denver to Golden, through Golden Gate Canyon, and into

FIGURE 2.1. Eben Smith, ca. 1875. History Colorado, Object #89.451.4395.

Gregory Diggings and took seven to eight hours (far better than the three to four days the journey had taken the year before). Once the stamp mill was up and running, Smith and his family took the stage back to Denver and briefly went to Elwood, Kansas, just across the Missouri River from St. Joseph, but they soon returned to Central City.[5]

With most of the gold around Gregory Diggings trapped in quartz formations, stamp mills were a welcome addition to the mining activities there. While several stamp mills were built in the mining camps at about the same time, throughout Eben Smith's life the Smith and Chaffee Stamp Mill was frequently referred to as the first, and no one discounted the impact it had on mining. From the moment it opened, it was very busy processing ore from the area. Smith managed the mill; at that time he was the only man in Colorado who had the experience in the modern process of milling for the extraction of gold. Chaffee settled in Denver and paid very little attention to the mill, focusing instead on the financial aspects of their business dealings and leaving the mill to his partner's more "expert" knowledge. The *Rocky*

Mountain News wrote in 1862 that Smith and Chaffee retorted "the purest and nicest gold of any mill men in the mountains." The mill was so successful that at the time of Chaffee's death in 1886, the *Denver Republican* said it had "materially assisted in the development of the mines of Gilpin County. It gave the miners renewed hope, and added a great impetus to the industry."[6]

While milling was profitable, Smith and Chaffee also engaged in mining, buying several lodes in the area—many of which kept their mill running. They bought two claims on the famed Bobtail Lode, which produced nearly $3,000 a week in ore at times, and further developed John Gregory's original mining claim on the Gregory Lode. Their investment in the mill paid off handsomely in 1863, when they sold it for $250,000, which Smith's stepson Fletcher Jordan later described as a large profit. After selling the mill, the two men continued to work the Bobtail and Gregory Lodes for about a year, then they sold them as well. Smith and Chaffee later repurchased the properties and consolidated them into what became the Bobtail Lode and Tunnel. The two began shifting their attention from gold to silver because, as Smith later explained, they thought it was more profitable than gold mining at that time.[7]

Just as miners had done in California, the miners arriving in Central City and the other mining camps had to create a new society. While Smith had played a very small part in establishing a new society in California, he played a much larger role in Colorado socially, economically, and politically. If helping to establish that new society also let him gamble with some new business ventures, he was happy to help. In 1864 Smith bought the Montana Theater, located at the head of Main Street in Central City, but he owned it for just one year, selling it to the Langrishe, Barnes and Jones Theater Company (headed by famed theater man Jack Langrishe). He also became involved in the banking business, which proved more profitable than theater. In 1865 Smith, Chaffee, and several others bought the banking house of Clark, Gruber and Company in Denver, which they reorganized as the First National Bank of Denver. Smith was wealthy enough at that point to finance construction of the bank's new home, the National Block, located at F (now Fifteenth) and Blake Streets in Denver. Built at a cost of $45,000, the building housed the First National Bank and three storefronts. Smith originally leased the building to the bank for $2,500 a year, but the board of directors purchased it outright in 1866.[8]

His involvement with the First National Bank also introduced Smith to a new associate in his mining ventures, David Moffat, a clerk at Clark, Gruber and Company. Born in Orange County, New York, on July 22, 1839, by age twelve Moffat was working as a messenger at the New York Exchange Bank. By age sixteen he was an assistant cashier, at which time an older brother working at a bank in Des Moines, Iowa, urged David to come there. He did and found work as a messenger on the stage line that ran from Des Moines to Omaha, Nebraska. In Omaha, he caught the attention of Benjamin F. Allen, a local banker, who in 1856 made Moffat his bank's cashier. The bank failed in 1859, and Moffat was responsible for winding up its affairs, making sure to pay every dollar owed to depositors.[9]

Moffat bought and sold lots in Omaha for a short time after the bank's failure, although he had to transact business through other people since he was not of legal age. The news of gold in Colorado attracted his attention, and he formed a partnership with C. C. Woolworth to set up a book and stationery store in Denver to supply the miners passing through town. Moffat ran the store in Denver while Woolworth forwarded supplies as needed. The store also housed a post office, and Moffat served as assistant postmaster. His reputation grew, and during the Civil War, he held the position of adjutant general of Colorado under territorial governor John Evans. In 1867 he became cashier of the First National Bank of Denver, another step on his way to the eventual presidency of the bank. He also became a permanent fixture in the Smith and Chaffee ventures, though he was only a minor figure at first.[10]

Eben Smith was active in Colorado politics in one way or another during most of his years in the state. In March 1861, less than a year after he arrived in Central City, organizers held a meeting aimed at combining the Lake Gulch and Quincy mining districts. Smith, who favored the combination, attended the meeting and, as an eligible voter, served on a committee that demanded the books from the officers of the Lake Gulch and Quincy districts. When William Gilpin was forced out as territorial governor in 1862 as a result of unauthorized drafts he issued on the federal government to help raise troops to protect the Colorado goldfields from Confederates during the Civil War, several people began a movement to strip his name from the county Eben Smith called home. Smith and Henry Teller were among those who opposed the movement and successfully argued to preserve the county's name.[11]

Smith's political career truly took off in 1865 when in September, voters chose him as Gilpin County's representative to the next Territorial Council, the equivalent of today's state senate. The next month, Gilpin County voters chose Smith as their delegate to the Union State Convention, the purpose of which was to reaffirm Colorado Territory's support for the government of the United States. The fact that Smith was chairman of the Anti State Central Committee at a time when most residents of Colorado were pushing for statehood did not deter the county's voters from choosing him to represent them. One issue that still polarized the territory at the time of the convention was the massacre of more than 150 Cheyenne Indians, mostly women and children, at Sand Creek in southeastern Colorado the previous November. At the time of the massacre, the Indians believed they were under the protection of US authorities at nearby Fort Lyons, but that did not deter members of the Colorado Volunteer Cavalry under the command of Colonel John Chivington from attacking them. Delegates to the convention adopted eighteen resolutions by majority vote. Four of those resolutions, known as the Sand Creek Platform, condemned public attacks on soldiers who had taken part in the massacre and called for more attacks against the Indians.[12]

Eighteen of the seventy-three delegates, including Smith, voted against the Sand Creek Platform. The Central City *Daily Miner's Register* suggested that many of the delegates who voted against it did so out of concern that popular support for the Sand Creek Platform might prevent Colorado from entering the Union. According to the *Rocky Mountain News*, the resolutions were a "true index of the popular feeling," and those opposed to them would find "if they take half the pains to ascertain the will of the people that they have taken to choke it down, that the masses are packed against them." Smith's leadership of the Anti State Central Committee seems to suggest that he was not concerned with statehood, further suggesting that he may have sympathized with the victims of the massacre.[13]

On January 2, 1866, three months after the convention, Smith traveled to Golden to take his seat on the Territorial Council. One of the first votes he cast was to move the council meetings to Central City, but he lost out to the majority, who voted to adjourn to Denver. In the council, he sat on four committees—Military Affairs, Highways and Bridges, Indian Affairs, and Finance, Ways, and Means (of which he was chairman). One of the biggest controversies facing the council during that session was the Sand Creek Massacre. Territorial governor Alexander Cummings was deeply troubled by the division in public opinion over the massacre and the role it had played in the Union State Convention. Cummings did not want it to become a similarly divisive issue during the legislative session, but it continued to dominate the political scene for months to come.[14]

Smith introduced three bills during the January–February 1866 session of the council. The first was a bill concerning the drainage of mines, which was introduced on January 17 and passed on January 26 but ran into problems when it was sent to the Territorial House for its approval. The second bill dealt with refining the definition of weights and measures, which passed on February 3. After some minor changes, the house passed the bill and the governor signed it. Smith also introduced a bill to hold a supreme court in the territory and assign districts to judges, which also met with the eventual approval of the house and governor. Overall, however, Smith was relatively quiet during his time on the council. In fact, his most heated debate concerned setting the salaries for judges. The council adjourned in mid-February and reconvened in December 1866, to run through January 1867. Smith served on the same committees, but he was absent from the sixth session of the council a great deal more than he had been during the fifth session, most likely due to business and personal concerns.[15]

One new business venture came Smith's way as a result of action by the council during the sixth session. One of the bills passed incorporated the Georgetown and Breckenridge Wagon Road, which connected the two cities. The bill allowed builders of the road to erect three tollgates, two near Breckenridge and one near Georgetown. The legislature gave the contract for the road to Joseph Watson (who became a longtime friend and associate of Smith and his family), William C.M. Jones, Richard Irwin, William A. Hamill, George W. Clark, and Eben Smith. He did not vote on this bill in the council, so he may have excused himself to avoid a conflict of interest. The road was commonly known as the Watson road, and it was still in operation (though in poor condition) in 1890 when Clear Creek County purchased it and finally eliminated the toll on it.[16]

While Smith's business life was booming, his personal life was busting. In 1863, Eben and Caroline had a daughter, Nellie, but the next year their marriage fell apart and they divorced. Smith's stepson Fletcher Jordan blamed the divorce on "the other woman," Smith's second wife, but there is no indication that he met that woman until well after his divorce was final. In 1864, Caroline and her seven children moved to Galesburg, Illinois, where the Smiths had spent some time before coming to Colorado. Eben bought Caroline a house there and transferred $25,000 in government bonds to her. The interest from these bonds "afforded her a comfortable living for herself and family," and the family stayed in Galesburg while the children were still in school. Caroline eventually returned to California and settled in Oakland, where she died on May 22, 1891, at age sixty-five. Following Caroline's departure from Colorado, Smith's sister-in-law Jennie Smith (wife of his brother Samuel) moved in with him to help keep house. Smith picked Jennie up in St. Joseph on his return from New York, where he and a group of other Coloradoans had entertained former territorial governor William Gilpin. That same year, Eben's brother J. Nelson, who was second major of the Second Colorado Cavalry, was killed in the Battle of the Little Blue on October 21.[17]

No matter what was going on in his personal life, business was never far from Smith's mind. On January 9, 1866, the owners of the First National Bank unanimously reelected the entire board of directors, including Smith. His time with the bank would end in July, though, when he sold all of his stock and resigned from the board. He was also able to pay attention to all of his mining ventures without seeming to lose track of any of them, a trait he would demonstrate throughout his life. While politics and his new business ventures were enjoyable and profitable, Smith's true love was mining; as soon as the second session of the council adjourned, he gave it his full attention.[18]

By 1866, stamp mills were no longer as effective in processing Gilpin County ore as they once had been, as the rock formations deeper in the mines produced low-grade sulfide ores that required more sophisticated processing. Miners throughout the county attempted to develop new ways to treat the ore, including a refined stamp mill that Peter McFarlane helped develop. Nathaniel Hill, a professor of chemistry at Brown University in Providence, Rhode Island, first arrived in Colorado in 1864 to investigate mines in and around Gilpin County. By 1865 he was studying smelting methods, and in 1866 he traveled to Swansea. Hill opened his Boston and Colorado Smelter in Black Hawk in 1867, and the mining industry in Gilpin County recovered from the depression that had plagued it the previous two years. Smelting, as developed by Hill, involved first roasting the ore and tailings to reduce the sulphur content, which was high in Gilpin County's ores. Roasting produced a concentrate of gold, silver, and copper that could then be refined. Smelters continually refined the process in the years after Hill opened his first smelter, all with the goal of recovering more valuable minerals from ore.[19]

Smith and Chaffee made their own first attempts at smelting when they built the Narranganset Works on the Gregory Lode in February 1866. The Smith-Chaffee Narranganset Company owned 400 feet of the lode, and one of the shafts reached a depth of 350 feet. Newspaper reporters from the *Rocky Mountain News* toured the mine, and what they saw amazed them. As the reporters wrote of their descent into the mine, they described the different tunnels that led from the main shaft and reported that the miners had found "a vein of very fine" gold ore. A second shaft off the main one had also opened a "valuable source of ore," leading the reporters to write that the mine was "now in the best possible condition for taking out large amounts of very valuable ores." The reporters thought the Narranganset Works were "the model of the mountains" and that Superintendent Eben Smith "may well feel proud of the thorough, systematic arrangement he has perfected." In two weeks in March 1866, the Narranganset Works had paid out $7,500 worth of gold. Smith, more modest than the *News*, gave credit to Nathaniel Keith's desulphurizing process combined with James Lyon's smelting process for the mine's success.[20]

Smith and his sister-in-law Jennie entertained William Newton Byers and others on the editorial staff of the *Rocky Mountain News* after their inspection of the Gregory and other Central City mining operations, and Byers wrote that this helped make their visit to Central City "agreeable and happy." At about that same time, Smith was preparing to marry Emily Louise Rundel, a twenty-nine-year-old teacher at a one-room school in Boulder. Emily was born in Rochester, New York, on May 13, 1836. In 1860, she and her brother Giles set out for Colorado, eventually settling in Boulder. With his approaching marriage, Eben no longer needed Jennie Smith's help, and on March 6, 1866, he accompanied her as she traveled to Pennsylvania. Smith did not return to Colorado until early August, but even in his absence his reputation grew, and when the *Rocky Mountain News* announced his return it referred to him as one of "Colorado's headest men." Eben Smith and Emily Rundel

FIGURE 2.2. Emily Smith, early 1890s. Louis O'Kane Collection.

were married in a log cabin in Boulder on September 23, 1866, and newspapers described their wedding as one of the social events of the year. The Smiths moved into a new Gothic-style house Eben had built on Casey Avenue in Central City. Their daughter, Kate May, was born in July 1867.[21]

By the time of his marriage, Smith's mining operations had expanded to Georgetown, in part because of the difficulty still plaguing ore processing in Gilpin County's mines. There, he and Chaffee invested in Dr. Charles Johnson's Georgetown Smelting Works, which was able to take 8 ounces of ore and process 4 ounces of silver from it. The Georgetown mines were producing at least a ton of silver a month according to some, and the *Rocky Mountain News* reported that "no one, except he that has seen a test of the ores from the mines in the vicinity of Georgetown, can possibly believe in their astounding richness." In June 1867, Smith, Chaffee, and Fox Diefendorfer bought 1,400 feet of the Anglo Saxon Lode, located near Georgetown, from J. T. Harris, T. J. Campbell, and Dr. Darnell for $50,000 "cash in hand." In

FIGURE 2.3. The Smiths' house on Casey Avenue in Central City, 1942. Denver Public Library, Western History Collection, X-2704.

addition to mining, Smith and his associates also built a smelter on the property. In November 1867, a reporter from the *Rocky Mountain News* saw 97 ounces of ore from the Georgetown mines in Smith's hands, an impressive sight. Smith also reported that his mill was only working half time due to a lack of water.[22]

In July 1867, Smith's supporters re-nominated him as a candidate for the Territorial Council, and the *Daily Miner's Register* endorsed him. He lost to a fellow Republican, coming in sixth with 836 votes. What role his controversial vote on the Sand Creek Platform at the Union State Convention the previous October played in his defeat is impossible to determine. The *Miner's Register*, however, viewed the overall results of the election as a stunning defeat for Republicans. His loss did not slow him down, and in August 1867, he served as a horse judge at the Second Annual Exhibition of the Agricultural Society of Colorado. Freed from political cares, Smith continued to prosper at mining. The down time at the mill left Smith feeling bored, which he did not like, and in his continuing search for profit he diverted himself with a new business venture. In January 1868, Smith became involved with the construction of his first railroad, the Rocky Mountain Railway and Telegraph Company. Smith was a director and incorporator of the new line, which was to run

FIGURE 2.4. Cora Smith, 1880s. Louis O'Kane Collection.

from Denver to the coal beds of South Boulder, then to Black Hawk, Central City, and Georgetown via the Ralston Creek route. If completed, the railroad would have been a great benefit to the Smith-Chaffee mining operations, but like several of Smith's business ventures, it never amounted to anything.[23]

While his business life prospered, his personal life experienced another big blow. There would be a number of tragedies associated with the Smith family over the years, and the first one struck in 1868. On August 24, Kate May died at age thirteen months. Indicating how important and influential Eben had become, Kate's lengthy obituary in the *Central City Register* told of "another little angel" that "hath taken to itself wings and flown away to the bright and shining shore." Other obituaries for children in Central City at this time were not nearly as emotional. Two days after her death, the Smiths held Kate's funeral at the house on Casey Avenue. The following year, on March 27, 1869, Eben and Emily had their second child, a daughter named Cora Isabel. Their son, Frank Leroy, was born on May 26, 1871. In August

1876, the family narrowly avoided another tragedy when five-year-old Frank, while trying to jump onto a wagon on Eureka Street in Central City, was run over by the wagon. Fortunately, the wagon was empty, and the injuries to the "little fellow" were not serious.[24]

By the time of Frank's birth, Chaffee's advancing political career left Smith with almost all of the responsibility of running the Smith-Chaffee operations. In 1865, when Colorado was attempting to become a state, voters chose Chaffee and territorial governor John Evans as the hoped-for new state's senators. However, Colorado failed to gain admittance as a state when President Andrew Johnson twice vetoed the bill due to Chaffee and Evans's failure to support his Reconstruction program. The people of Colorado wanted Chaffee in Washington, though, and in 1870 they elected him the Colorado Territory's delegate to the US Congress and reelected him in 1872. Smith, a member of the Republican Party since its founding, used all of his influence to help the cause of his friend and business partner. Chaffee's chief goal in Congress was to get Colorado admitted as a state, which he worked at "continuously and persistently." Whether he ever discussed Smith's membership in the Anti State Central Committee is unknown.[25]

By 1873, one year before the value of silver produced in Colorado exceeded that of gold, Smith increasingly focused his attention on silver mines located around Lincoln and Bross Mountains near Alma. Dozens of old miners from Gilpin, Clear Creek, and Boulder Counties crowded the Denver, South Park and Pacific Railway line in the summer of 1873 as they made their way to these mountains to try their luck. Much of the ore was in a lime formation and was thought to be very rich near the surface based on discoveries already made. The town of Quartzville, located at the foot of Lincoln and Bross Mountains, was the closest town to the mines; the *Rocky Mountain News* reported that about 100 houses, stores, and a saloon sprang up there during the 1872 season. The town also had its own steam crushing and sampling works, built by Jerome Chaffee, Eben Smith, and Dr. John H. Morrison. Professor Eugene Burlingame, former Colorado territorial assayer, ran the works, which were described as "a grand thing for the miners" since they were able to get their ore crushed and weighed so close to the mines.[26]

Smith and Chaffee turned their attention to silver mining just as the "Crime of 73" was taking place. For many years the federal government had maintained a ratio of 16 to 1 with gold and silver, with 16 ounces of silver equaling 1 ounce of gold. As such, the federal government bought silver at $1.20 an ounce to use for silver coins. The price on the international market was usually higher. In 1853 the federal government stopped using silver for all but dollar coins and then dropped the silver dollars in 1873. At the same time, silver production in the United States was at an all-time high, thanks in part to work at the Comstock Lode in Nevada and Colorado's silver mines. Silver miners were not particularly concerned at the time because of the international market, but when several countries dropped silver in favor of gold, the miners turned to the federal government, only to find it was no longer buying silver. Western congressmen joined with representatives from farm states to pass the Bland-Allison Act in 1878, which required the federal government to buy a minimum of $2 million in silver each year. It was not the 16 to 1 of old, but it was something. It was the start of a long fight on behalf of silver in the United States, and Colorado's miners could count on Senator Henry Teller as one of their main supporters.[27]

Smith returned to politics on February 5, 1874, when he became the seventh postmaster of Central City, replacing his friend James D. Wood. He served in that position until December 4, 1876, when he gave it up as he prepared to follow the Smith-Chaffee mining operations to Boulder. When Colorado became a state in 1876, Jerome Chaffee finally made it to the US Senate, though he served only a short term so elections could remain staggered. Central City lawyer Henry Teller, the other new senator, served the usual six-year term. Despite his shortened time in the senate, Chaffee "devoted his abilities and spent his money liberally in advancing the interests of the State."[28]

In mid-1877, the Smith family left Central City. During Eben Smith's seventeen years there, he continually helped build up the town and the state through his work as a smelter owner, mine owner and manager, railroad builder, theater owner, and politician. He also took great joy in more lighthearted activities such as hosting the editor of the *Rocky Mountain News* and judging horses at the agricultural exhibition. Although Smith would continue to keep up a pace in the future that would make many people dizzy, he was never as active civically as he was during his time in Central City. Many men in his financial position felt they owed something to the community in which they lived, which likely accounted for many of his social and political activities. It was also logical for his fellow citizens to call on him, as a respected member of the community, to help stabilize the state. However, new challenges awaited Smith in Boulder, and he was eager to take them on.

Samuel Conger, William Martin, George Lytle, Hugh McCammon, John Pickel, and Samuel and Harvey Mishler discovered the Caribou Mine near Nederland in 1869. The group tried to keep their discovery secret, but before long word got out, and by June 1870, miners inundated Caribou. In a short time, a town appeared "as if by magic" on the site of the original seven men's cabins. That summer, miners discovered a number of other important mines in the area, including the Comstock, Sherman, Spencer, and Idaho Mines. The amount of ore taken out of the Caribou created a great deal of interest among speculators, and in September 1870, Abel Breed and Benjamin Cutter bought the western half of the property for $50,000. The two men bonded the eastern half for $75,000, but the Caribou Company retained ownership of it. Breed and Cutter further developed the western half of the mine, and between October and December 1870, they sold 88 tons of ore to the Boston and Colorado Smelting Works in Black Hawk for $9,000.[29]

In 1872, with the mine prospering, Breed decided to build a mill to handle his ore onsite rather than ship it to Black Hawk or Central City. By the spring of 1873 the mine was doing so well that Breed decided to sell it to a Dutch company, Mining Company Nederland, for $3 million. The new owners experienced a few difficulties, in part because they spent all of their available money just to purchase the property but mostly because Breed had stripped the mine of all available ore when he knew the sale was going to take place. Mining Company Nederland said it expected to find paying ore by year's end, but the "rascality and incapacity" of the "kid-gloved gentry" in charge led to more problems, and the mine closed in December 1875.[30]

As soon as the mine closed, miners and creditors filed 150 liens against the property, including one by Jerome Chaffee for $47,933. By March 1876, Mining Company Nederland was bankrupt and the company's assets, including the Caribou Mine and the mill, went up for auction. Chaffee and David Moffat won with a bid of $70,000. The fact that Chaffee had been involved in the original lawsuits that forced the sale of the property raised questions, and some accused him of butchering the mine and attempting to demoralize the Dutch stockholders. There were also rumors that he had been involved in Breed's earlier sale of the mine for more than it was worth. Chaffee denied the charges but never completely cleared himself of suspicion. After buying the mine, Chaffee and Moffat immediately appointed Eben Smith superintendent.[31]

The Caribou was never completely inactive during the financial troubles that plagued Mining Company Nederland, but when Chaffee, Moffat, and Smith took over, they had to do a great deal of work. The biggest problem in need of an immediate solution was that of rising water, very common in the mining business. Shortly after taking over, Smith wrote to Moffat that the bottom of the main shaft looked promising for ore, but the vein was not yet settled and from what he could see, the rock looked to be low grade. In 1877, miners under Smith started taking ore out of the mine, and Chaffee reopened the mill.[32]

As operations got under way, Chaffee did what he could to clear up the final liens against the property with the money earned from the mine during the winter of 1875–1876. Smith and Chaffee's efforts met with success, and by 1878 the mine looked to be prosperous for the first time in many years. The next year, the mine had become a "steady producer." Workers sunk the main shaft, one of seven total, to a depth of 740 feet. There were thirteen levels to the mine, all of which were connected by winzes. A two-compartment bucket operated in the shaft, and a Knowles double-action pump helped drain the water from the mine. The hoisting house enclosed two shafts, a blacksmith shop, an air compressor, sorting floor, tramway cars, and the offices. There were also buildings and machinery over two of the other shafts. With the mine in good shape, Chaffee and Moffat decided that the best way to make a profit was to do as Breed had done and sell the mine. In April 1879 they sold the mine for $900,000 and the mill for $100,000 to the Caribou Consolidated Mining Company of New York. Eben Smith stayed on as general manager.[33]

On September 14, 1879, the operations at the Caribou suffered a severe setback when a fire destroyed the main shaft house and the machinery inside it. Although there was insurance on the property, the $10,000 it paid was not nearly enough to cover the damage. Smith was a fiercely determined man, described by some as "iron nerved," and he did not let this setback discourage him. He immediately started rebuilding the shaft house and replacing the destroyed machinery, but the stock of the mine dropped from $6.50 to $4.25 a share. Still, Smith moved quickly, and when he paid a visit to Central City in December, he reported that the burned buildings had been rebuilt, including a new shaft house at a cost of $35,000, and that he expected to resume paying dividends in early 1880.[34]

While visiting Denver in mid-December, Smith reported that he expected to have the new machinery operational soon. The fact that Smith was able to be in Denver at all was a result of the good luck that had followed him to that point in his mining career. Just a few days earlier, Smith and an assistant, Richard Crow, had set out for the mine in a sleigh during a blinding snowstorm. About a half mile from the mine, Crow lost control of the horses; and Smith, Crow, and the sleigh were thrown off the edge of the road into a snowbank almost 20 feet below. According to the *Rocky Mountain News*, Smith, "an old Coloradoan" with "extended experience" in the mountains, knew that the men had to move quickly to survive. After getting the horses and sleigh back on the road, Smith had to make three more trips to get the packages and wraps. He collapsed into the sleigh after the third trip. The two finally made it to the Caribou, where it took Crow nearly an hour to recover to the point where he could speak.[35]

By the end of the year, the situation at the Caribou had gone from bad to worse. Robert Dun, who was the president of Dun and Company (later known as Dun and Bradstreet), claimed that the Caribou shaft had encroached onto his adjoining mining property. A series of lawsuits followed, with Dun and the Caribou owners suing each other. Gilbert Lehmer, another area mine owner, filed his own lawsuit against the Caribou, claiming that Caribou miners were stealing his ore. The Colorado Supreme Court eventually dismissed Lehmer's case, but the trouble with Dun was more difficult to sort out. Smith was especially anxious for the two sides to reach a settlement, as he thought the Caribou would soon go broke if the situation continued as it was. In early 1880, rumors of a merger between Caribou Consolidated and Dun's mining properties started to spread, and officers of the Caribou Company met with Dun and his attorney, future US president Chester A. Arthur, to talk about consolidation. After intense negotiations, on June 25, 1880, the two sides agreed to merge. The Caribou and Dun's No Name, Spencer, and Columbia Mines became one, although the new company kept the Caribou Consolidated name. Dun believed he had a good manager in Eben Smith, and despite the trouble between the two companies before the merger, he hired Smith as superintendent and general manager.[36]

FIGURE 2.5. Abel Breed, William Bryant, and Eben Smith (*left to right*) examining ore from the Caribou. Denver Public Library, Western History Collection, K-323.

Under Dun's ownership and Smith's management, the mine prospered. By May 1881, Smith was employing eighty-three men at the Caribou, and they were digging ore on the 860-foot level of the mine. Smith kept the mill at the Caribou running at full capacity nearly every day, processing ore from the Caribou and No Name Mines. The Caribou, in fact, was producing more ore than the existing mill could handle, and Smith was busy designing improvements that would allow the mill to increase capacity. Overall, the mine was a "grand success" due to the "business-like" way Smith, who was "well known as one of the best mining men in Colorado," had been running it for the last five years.[37]

Work at the mine was Smith's passion, but shortly after he arrived in Boulder County, he also resumed his political career. In 1877, Governor John Routt appointed him a Boulder County commissioner, a position to which he was reelected in 1878. He served as chairman of the Board of County Commissioners during his second term. Smith's two terms as a county commissioner were fairly quiet, as he dealt mainly with routine issues. One example was the procurement of a courthouse for Boulder County. On October 27, 1879, the board considered a proposal from Charles Van Fleet to rent or lease Van Fleet Hall to Boulder County as a courthouse until May 1, 1880, and then for two years after that at whatever price the county commissioners decided on. Smith moved that the county rent Van Fleet Hall for $1 until May 1 and for two years after that at $200 per year. The other commissioners accepted his plan, and Boulder County found itself with a courthouse.[38]

In late 1880, Smith declined to run for a third term as commissioner, in large part because a law recently passed by the Colorado State Legislature required all county commissioners to post a bond of $25,000 or face a $500 fine every time they tried to act as a commissioner. The *Rocky Mountain News* argued that the legislature had passed this law because of pressure from people in Leadville, where the commissioners were "objectionable to the politicians," and referred to the men of the Colorado General Assembly as "these smart Alecks" for passing it. It went on to report that J. B. Thomson, the Boulder County commissioner for Longmont, wrote to Smith that he had received his bond for his signature "and that he had filed it—in the stove." Smith said he "did not feel like giving such an enormous bond," and the *Rocky Mountain News* feared that on July 1, 1881, there would be no legal county commissioners left in the state. This was the last elected office Smith held, but it was not the end of his involvement in politics.[39]

Smith's personal life reflected his prosperous mining career. In July 1880, he moved from Nederland to Boulder, where he bought a house at 1902 Walnut, which he immediately had extensively remodeled. The *Boulder County Herald* commented that "when finished it will be a handsome place." Twenty-two-year old Lemuel Smith rejoined his father around this time and found work as an assayer, most likely at his father's mine. But by 1881, the fifty-year-old Smith was feeling restless. He decided that five years at the Caribou was enough and left his job as superintendent. Freed from the worries of managing a mine, he went on to help incorporate the Boulder, Middle Park and Grand River Railroad Company while still living in Boulder. The new railroad, Smith's second, went from Boulder to Sunshine Canyon, then to Nederland and Middle Park and finally into Routt County. Many people thought nothing would come of the railroad when it was in the planning stages, but as construction progressed, that sentiment changed. An article in the *Rocky Mountain News* said that the railroad would make Boulder "second to Golden in the possession of smelters" and that it "would be of more importance to this city than any road that could possibly be built."[40]

While work on the railroad progressed, Smith lost his substantial fortune. A letter written in 1902 by longtime friend and business associate David Moffat suggests that this was a result of stock market losses. At that time, Smith was organizing a stock deal in New York, and Moffat warned him, "I should think that your past early life would have taught you that [that] was another man's game you are playing in," adding that he hoped Eben would "get out alive." Despite this setback, though, he was still a respected mining authority in Colorado, with a number of mining successes to back up his favorable reputation. During his five years in Boulder, he had managed to turn the nearly defunct Caribou Mine into a successful producer. His valuable experience at draining the water from the mine would soon come in handy as he pursued a new venture. Smith was not as active civically in Boulder as he had been in Central City, but his two terms as a county commissioner showed that his fellow politicians and voters still thought highly of him. As H. W. Brands writes, miners who built their fortunes in California were used to also losing them. He explains, "A person was expected to gamble, and to fail, and to gamble again and again, till success finally came." In an environment where failure was common, failing lost its sting and no one counted the losses, only the successes. Determined to rebuild his fortune, Smith left Boulder in 1882 with $5,000 from the sale of his house and the prospect of a new job working for Horace Tabor.[41]

3

"David Moffat Set You on Your Feet"

Horace Tabor and Jerome Chaffee joined forces in 1881 to buy the Tam O'Shanter group of mines—including the Montezuma—in Ashcroft, near Aspen. Before finalizing the purchase, Tabor hired a number of mining experts to inspect the property. After all the experts reported back favorably, he agreed to the $100,000 price, although the contract stated that he could put $5,000 down and pay the remaining $95,000 in ninety days if the property justified the purchase. Tabor's workers soon discovered a rich silver vein, and a thrilled Tabor finalized the deal in September 1881. A town quickly grew up around the mines, and when Tabor announced that he and James B. Grant, a well-known smelter man from Leadville, planned to build a smelter to process the ore, the value of the mine increased dramatically. Tabor further increased the value when he announced that he planned to start building a road to the mine.[1]

Unfortunately for Tabor, Samuel Bruckman, a local mining and real estate broker, filed a lawsuit against the Tam O'Shanter in January 1882. Bruckman sued for what he said was his promised share of the mine for financing its

DOI: 10.5876/9781646421794.c003

original discoverers, which tied up Tabor's purchase of the property. Tabor, seeking to protect the $95,000 he still owed on the mine, put it into escrow. Then, the original discoverers of the mine, James Chaney and Nicholas Atkinson, filed their own lawsuit to get the mine back. The Colorado Supreme Court finally decided the issue in favor of Tabor in 1888, but by then the mine was virtually worthless.[2]

While the courts sorted out the lawsuits, Tabor continued to work the Tam O'Shanter. In mid-August 1882, he arrived in Ashcroft accompanied by his latest expert, Eben Smith, for another inspection. The *Rocky Mountain Sun* reported on August 19 that the "testimony was in" and Tabor "ran no risk in developing the mine." Tabor hired Smith as general manager, and by the end of August, Smith was building a house for himself and his family in Ashcroft and had given W. H. Coxhead a contract to complete a road to the Tam O'Shanter within thirty days.[3]

By the beginning of September, the road crew consisted of forty men working full-time on the road to the Tam O'Shanter, and Smith had a full force of experienced miners at the mine, which prompted the *Sun* to report that "things are looking better." Smith and his family were in their new house by October 21, and he was getting the mine ready to work through the coming winter. Smith had new bunkhouses and boardinghouses built and ordered five carloads of provisions and 300 cords of wood from Denver. Smith employed a total of twenty-five men at the mine during the winter of 1882–1883, and the *Leadville Daily Herald* wrote, "we predict that under the able management of Prof. Eben Smith these mines will show something wonderful by next spring." The *Herald*'s prediction seemed to be coming true, as at the beginning of December there were 2,000 tons of ore on the dump at the Tam O'Shanter waiting to be processed. Smith said all Ashcroft needed was a smelter, and "if one is started he will supply all the ore necessary to run it."[4]

Unfortunately for Smith, appearances were deceiving. In addition to trouble with actually getting ore out of the mine, non-Cornish miners struck against a Cornish foreman in the spring of 1883. Whether Tabor dismissed Smith or he left on his own is not clear, but after less than a year, he was gone from the Tam O'Shanter, replaced by John Tabor. Work on the Tam O'Shanter stopped in 1884 as Tabor, too, was unsuccessful. What no one seemed to realize, or at least refused to tell Tabor, was that the ore vein was actually very shallow and not very valuable.[5]

After his failure at the Tam O'Shanter, Smith left Ashcroft and went to Red Cliff, a five-year-old silver mining camp that was flourishing. Smith spent a few unsuccessful months there before David Moffat came to his rescue, hiring him as general manager of his mining interests in Leadville. Smith's long list of technological achievements would grow as he confronted the wet mines of the Leadville district, in the process restoring and increasing his lost fortune. The move ruined him in the eyes of some, such as a miner named Osborn, who in 1906 wrote to Smith that he had been a "good man once, but David Moffat set you on your feet."[6]

Moffat biographer Steven F. Mehls writes that Eben Smith was "one of Colorado's most astute mining engineers." Although he did not have formal training, he had a "sixth sense as to which mines would be profitable and which would not," and he could run a mine any way the owner wanted it run. The partnership between Smith and his skill with a mine and Moffat and his financial resources that began in Leadville in 1884 "formed an unbeatable team." Smith's role as a mining engineer was increasingly in demand beginning in the 1880s, as Gene Gressley writes that mining engineers "became more and more relied upon to guide their investors' inclinations." Within ten years of joining forces with Moffat, Eben Smith would be one of the most respected mining engineers in Colorado.[7]

The Leadville in which Smith arrived was at the beginning of a major mining boom. Prospectors first searched for gold in the area in 1860 and 1861, but when the placer deposits ran out, so did most of the people. Most miners ignored the large silver deposits in the area due to the lack of smelters that could handle them. Interest in the silver increased in 1872, and a group of miners spent several years trying to convince smelters to invest in Leadville. By the time the Harrison Reduction Works opened in October 1877, the boom was already under way. The next year, more than 40,000 people arrived in Leadville hoping to make their fortunes. One of the most famous who did strike it rich was Horace Tabor, who grubstaked the miners who discovered the Little Pittsburgh Mine.[8]

John McCombe and William and Joseph Pierce discovered the Maid of Erin silver mine in 1877, and two years later McCombe sold half of his interest in the mine to Tabor, Chaffee, Moffat, and Senator James G. Blaine of Maine for $43,000. The same group also bought the Henriett and Louisville Mines, all of which O. A. Harker managed until Eben Smith arrived in 1884.

Smith moved his family into a house at 214 West Sixth, opened an office at 320 Harrison Avenue, and quickly went to work improving the mines. By July 1886, the ore shoots he had opened at the Louisville Mine were improving every day, and it was "one of the most promising in the Leadville district."[9]

After leaving the senate in 1879, Jerome Chaffee spent most of his time away from Colorado, although he did follow his mining interests. On February 9, 1886, Chaffee returned to Colorado to inspect the new strikes in the Maid and Henriett, and it was obvious that his health was failing. Smith and others tried to convince him to skip going to Leadville in an effort to spare his health, but he made the trip anyway. In Leadville, his condition worsened and Chaffee decided to return to New York to be with his daughter, who was married to Ulysses S. Grant Jr., and his grandchildren. He left Leadville on February 24, arriving in New York just a few days before his death on March 9 at age sixty. Grant was responsible for settling his father-in-law's estate, thought to be worth "millions," although no exact estimate was available as the true worth depended on income from the many mines in which he was invested. The son of the former president traveled to Leadville, accompanied by Eben Smith and David Moffat, as they helped him settle the estate of their longtime friend and business partner. After Chaffee's death, Moffat became Smith's chief business partner. David Moffat, according to his biographer Steven Mehls, saw mining as a means to an end. He was far more interested in being president of the First National Bank of Denver and building railroads, and mining helped provide money for both of those endeavors. While Smith's partnership with Moffat was not as equal the one he had with Chaffee, Mehls writes that by the late 1880s the Smith-Moffat partnership "represented the most practical experience and greatest financial capability yet concentrated in one combine and it would become a key factor in Colorado mineral exploitation for the next twenty odd years."[10]

In 1887, the *Carbonate Chronicle* reported that Smith had ordered a 150-horsepower hoister for the Maid of Erin and that the plant would be "one of the finest in the district when placed in working order." Later that year Tabor, Moffat, Jacob G.B. Dubois, and George W. Kassler reorganized the Maid and Henriett into the Henriett and Maid Consolidated Mining Company, with a capital stock of $1.5 million (divided into 150,000 shares worth $10 apiece). Smith was a director of the new company. One of the

FIGURE 3.1. The main buildings of the Maid of Erin Mine in Leadville. Denver Public Library, Western History Collection, X-60968.

biggest problems with the Leadville mines was water, but Smith—using his experience from the Caribou—handled it easily, equipping all of the Smith-Moffat mines with pumps. He kept those at the Maid of Erin, Henriett, and Louisville running constantly to prevent the water levels in the mines from rising. Smith ran into the same problem when he leased the Castle View Mine in 1888. To solve the problem at that mine, he drove a drift from it to the Maid and Henriett, whose larger pumps were capable of handling the water from the Castle View. The *Evening Chronicle* thought all three mines would likely become "large producers" under Smith's able management.[11]

Smith also took over management of the Wolftone Lease, which was near the Maid and Henriett properties. His management of this property showed an "unwearied persistence" and "implicit confidence in the capabilities and possibilities of a piece of ground." Smith spent nearly $250,000 on development and machinery, and the shaft reached a depth of 700 feet before the miners found the "best looking body of lead carbonate ore" ever found in the Wolftone workings. The results pleased Smith, seemingly proving to him that hard work and money could turn just about anything around—a belief that would come back to haunt him in the future.[12]

By 1889, local newspapers were calling Eben Smith one of Leadville's most "influential men," a testament to his success even at difficult properties like the Wolftone. As a sign of the high regard in which many in Leadville held him, that year they selected him as a delegate to a miner's meeting, the members of which wrote a report "praying" for restrictions on lead imported from Mexico to protect the market for the lead coming out of their own mines. In October of the same year Smith, long a faithful Republican, joined C. C. Parsons, J. Y. Marshall, and W. F. Patrick in their effort to start a new political party in Colorado. The men firmly believed political leaders in the state had become too corrupt, driving the state deep into debt and forcing it to raise taxes too much. To counter these evils, they formed the Independent Party of Lake County for the purpose of electing "good and pure men" and in 1890 nominated John G. Coy, a rancher from Larimer County, for governor. Coy faced Republican John Routt, Democrat Caldwell Yeaman, and Prohibitionist John A. Ellett. Routt beat his three opponents, and Smith, relieved that at least a Republican had won, gave up his brief flirtation with forming a new political party.[13]

Smith's return to wealth and prominence in Leadville also made his family a part of high society once again. Emily Smith often hosted events to raise money for causes she supported, such as the Leadville YMCA, but the family held just as many parties for other reasons. In July 1887, Emily hosted a reception for Josephine and Fanny Watson, the daughters of Joseph Watson, the former superintendent of the Morningstar Consolidated Mining Company in Leadville and Eben's old business partner from the Georgetown and Breckenridge Wagon Road. The two girls had been guests of the Smiths' daughter Cora for the previous three weeks. Society turned out "en masse" for the "elegant affair," with the gentlemen in dark suits and the ladies in "gowns of the latest fashions." New Year's Day was always a cause for celebration in Leadville, and for many years the tradition had been to hold an open house on that day. However, the practice of hosting a dinner party was quickly replacing that tradition, and on New Year's Day 1890 Eben and Emily, assisted by their daughter Cora, "gave a most delightful dinner party." Among those present was their future son-in-law Charles Carnahan, who was beginning his long pursuit of Cora. Eben's son Samuel was also with the family, arriving shortly after they moved to Leadville. Samuel, who suffered from a heart condition all his life, most likely worked for his father in some

FIGURE 3.2. Eben Smith, 1891. Gilpin Historical Society, Central City, CO.

capacity. Doctors had prescribed morphine to treat Samuel's heart condition, but on June 12, 1887, he died in Leadville from an accidental opium overdose at age twenty-eight. The Smiths buried Samuel in what would become the family plot at Mountain View Cemetery in Oakland, California.[14]

When Moffat became interested in several new mining ventures in Aspen in 1890, Smith sold his house in Leadville and moved to a new house at 320 West Main Street in Aspen. For the next several years, Smith split his time between Aspen and Leadville, and in 1891 he started a new mining venture in Leadville that would become one of his most profitable ventures ever. That year Smith, John F. Campion, A. V. Hunter, G. W. Trimble, and Charles Cavender formed the Ibex Mining Company in Leadville. The company's most important holding was the Little Jonny, on the northeast side of Breece Hill, but Ibex also owned the Uncle Sam, Archer, and Titan Lodes in addition to the older Glengary and Queen consolidations. Miners discovered the

Little Jonny in 1878, about the same time as the Matchless, Horace Tabor's famous mine. By the time Smith and his associates bought it and made it part of Ibex, they knew it held large deposits of gold ore. The men hired J. J. Brown, who worked for Smith as superintendent of the Maid and Henriett Mines, to be superintendent of the Ibex.[15]

Sinking a shaft on the Little Jonny was not an easy job, as it required going through several beds of quicksand and driving in heavy timbers endways to keep the shaft from collapsing. The entire process was very expensive, but it proved to be worth it as miners soon hit a valuable ore vein. By 1893, workers were shipping 150 tons of ore per day (holding about $25 per ton in gold) from the mine, and industry observers proclaimed that at that rate, it would produce $1,368,750 in gold a year, which would have been a "greater yield than that of any other gold property in the United States." The owners rewarded Brown for his efforts with 12,500 shares of stock of the 100,000 that were available, making the Brown family very wealthy. The Little Jonny soon became one of the richest gold mines in Lake County, and by 1916, the mine had a "marvelous record of millions in gold." As late as August 1898, Smith was still sending dividends from the mine to J. J.'s wife, Maggie (popularly known as Molly Brown).[16]

While not as profitable as the Ibex, Smith and Moffat's two major holdings in Aspen were just as important to Smith. One of these was the Franklin Mining Company, which operated the Franklin, Dr. Franklin, and Millinee Mines. Work at the Franklin began in 1868, with more than 300 tons of ore taken out by October of that year. The mine's prosperity declined over the following years, but in July 1885, the *News* reported that the owners would continue working the properties despite the difficulties they had encountered. Smith began receiving reports about the Franklin from Smith-Moffat associate D.R.C. Brown in 1887, but it was not until the 1890s that he became heavily involved with its management. Under the ownership of David Moffat and the management of Eben Smith, the mine improved. By 1892, the shaft at the Franklin was down 600 feet, and the mines produced 20 tons of "first class" ore per day. As had been the case with the mines in Leadville, water proved to be a problem in the Aspen mines. To handle the water issue, Smith and Moffat organized the Aspen Deep Mining and Drainage Company, their other big holding in Aspen. The aim of this company was to drain the Enterprise, Aspen Mining and Smelting Company and the Franklin, in addition to mining the Big Chief, Little Chief, and Homestake Mines. As late as 1901, operations continued in the Aspen Deep Mining and Drainage Company shaft, although the miners' progress slowed due to an "unusual flow of water."[17]

In addition to running their mines, Smith was also in charge of inspecting properties that interested Moffat. In March 1892, he wrote to Moffat in Denver that he believed Aspen would be a better camp than Leadville "ever was." He thought the particular property he was inspecting, owned by a man named Roeder, was a good investment, but he wanted to keep Moffat's involvement with it secret. He believed that if anyone learned of Moffat's interest, "every mother's son of them" who had bonds on the mine would want full payment, thereby driving up the price. From this point on in their business ventures, Smith tried hard to keep his and Moffat's involvement secret in an effort to keep prices down.[18]

The following May found Smith looking at a property owned by a man named Roudebush. On May 13, Smith wrote to Moffat that he had serious doubts about the property, and by May 14, his doubts had turned into a determination to have nothing to do with it. He was not shy about telling Moffat when he thought a property was bad, and he wrote to Moffat that whatever he did, "don't put me into this deal as I don't want any part of it." By this time, Smith and Moffat's reputation was so great that they received offers for dozens of mines from people who were sure they had found the next big one. Smith always investigated these offers, either in person or by reading assay reports, and he usually found that they were not quite as valuable as their owners believed.[19]

An example of one of these properties was the Holy Moses Mine, discovered in 1889 by Nicholas Creede in what would become Creede, Colorado. Seeking money for further development of the property, Creede offered Moffat the chance to invest, and Moffat and his partners put up $70,000 for further development. The mine prospered, although it did not get the attention from Smith that some would have liked. Charles P. Flinn wrote to Smith in December 1890 that "you have been so busy elsewhere that you could not give this property the attention you desire" and later stated that the mine was producing gold, silver, and lead. Early the next year Flinn wrote that he had great hopes for the Ethel vein of the mine, "but if it don't [*sic*] amount to anything, the sponge might as well be thrown up." Development at the mine continued into 1891, but Moffat may have sold his interest in it since Smith received no further reports about its condition.[20]

Eben Smith's duties required frequent travel between Aspen and Leadville, and in 1892 he bought a house at 120 West Seventh in Leadville. Both of his sons lived and worked in Leadville, though their occupations were quite different. Lemuel worked as a mining engineer. Frank, who moved into his father's new house, worked as a partner in the Reynolds Brothers' ore-hauling business. In a deal arranged by Eben, W. O. and Lincoln A. Reynolds agreed to sell Frank a one-third interest in their ore-hauling business for $2,250 in exchange for Eben giving them a long-term contract for hauling ore from the Franklin Mine and the Aspen Deep Mining and Drainage Company. Frank paid the $2,250 out of his profits at a rate of "several hundred dollars per month."[21]

Frank's involvement in Reynolds and Smith Ore Hauling and Heavy Teaming prompted someone to send an unsigned and undated letter, probably written in 1892, that came into Smith's possession. The unknown critic accused Smith of dishonest business dealings, such as having his companies pay to build his house in Aspen and diverting a car of coal from the mines to his house. The author also said that Frank's profits from Reynolds and Smith "were much larger than the investment would indicate," suggesting that bills were rendered for ore hauling that was never performed. In addition, the writer said that Frank had taken a salary of $125 a month from Eben's Aspen companies from December 1, 1890, to July 1, 1891, "for which he did not render any services other than entering the Company's office for not to exceed one hour on average of three times per week." Receipts in Eben's papers from this time refute the charge that he had his companies pay for his house, and it is difficult to imagine the very cost-conscious Smith paying bills for nonexistent ore hauling even if it was to assist his son. However, he had a soft spot for his son Frank, so it is conceivable that he would have done whatever he could to guarantee his success.[22]

Smith and Moffat became owners of two other mining companies in Leadville in 1892. They incorporated the Grey Eagle and Pocahontas Mining Company on July 11, 1892, although the two men actually bought the mine the previous September. In addition to the two properties that gave the company its name, the business owned the Crown Point, Bonus, and Little Rooster Mines. It also had leases on the Penrose, Orion, Starr and Alice

Lodes. The Grey Eagle and Pocahontas was the first deep-mining operation in Leadville, which miners had previously avoided because of the threat of water. Fortunately, according to some, there were men with "courage" such as Eben Smith and his brother Henry (later superintendent of the Grey Eagle and Pocahontas) who were willing to engage in these kinds of operations in Leadville. Henry W. Smith was born in Erie County, Pennsylvania, in 1848 and until the 1880s spent most of his life farming there. In an effort to make his own fortune, Henry joined Eben in Colorado in 1887 and started working at the Henriett and Louisville properties.[23]

The Grey Eagle and Pocahontas prospered under Henry's leadership. By August 1, 1892, the company had shipped 1,000 tons of carbonate ore and 2,500 tons of argentiferous ore, earning $48,447. Miners under Henry also uncovered a rich ore chute in the Penrose, which they predicted would yield 500 to 1,000 ounces of silver per ton. The Grey Eagle and Pocahontas was an impressive operation with a large hoist, three station pumps, and a condenser, though in late 1893 management turned them off as the water level in the mines receded. Encouraged by their success here, Smith and Moffat organized the Gazelle Mining Company on December 10, 1892, to work the Dillon, Elk, O. Z., and Bismarck Lodes. The Gazelle eventually became a subsidiary of the Grey Eagle and Pocahontas.[24]

Smith preferred to mill his own ores whenever possible, and in 1892 Smith and Moffat took charge of the abandoned La Plata Smelter in Leadville and formed the Bi-Metallic Smelter to treat sulphides and copper and iron pyrites. The smelter used William L. Austin's process, which required very little fuel to treat the ore, instead relying on the sulphides in the ore to create enough heat to complete the process. They were happy to process ore from other mines, but the main purpose of the Bi-Metallic was to mill the ore from the Smith-Moffat Leadville and Aspen mines. During the 1890s, Smith became increasingly distrustful of many of the smelters handling his ores, often feeling that they cheated him when it came to sampling and valuing the ore. Smith hired smelting expert Philip Argall to run the Bi-Metallic, but having his own smelter treat his own ores did not solve Eben Smith's troubles. For the rest of his life, Smith and his secretary were in almost constant disagreement with Argall over how valuable the ore actually was.[25]

With a few exceptions, Smith and Chaffee had all but ignored gold mining since 1864, believing silver was more profitable. As long as the US government coined silver, guaranteed by the Sherman Silver Purchase Act of 1890, which replaced the Bland-Allison Act, this idea held true. However, a British banking crisis in early 1893 led many British investors to sell their American holdings in exchange for gold, which, in turn, led to a drain on American gold reserves. By the time of Grover Cleveland's inauguration in March of that year, US gold reserves had fallen below $100 million, and Cleveland called for the repeal of the Sherman Act because it allowed for the redemption of silver certificates in gold. To the dismay of silver miners throughout the country, the US Congress repealed the Sherman Act on June 27, 1893, and within two days the price of silver dropped from 82 cents per ounce to 62 cents. This dealt a huge blow to Colorado's economy, and the silver-producing Smith-Moffat mines in Leadville felt the shock.[26]

In an interview with the Leadville *Evening Chronicle* on June 27, David Moffat said the properties under his control employed about 2,000 men who, in turn, were responsible for supporting five times as many people. Moffat said his loss caused by the repeal of the Sherman Silver Purchase Act was "trifling" when compared to the losses to his employees, and he went on to say that he hoped the workers would take a cut in pay since he could not afford to pay them $3 per day with the price of silver so low. The next day, as the crisis deepened, Smith closed all of the Smith-Moffat mines in Leadville. Numerous business failures and fluctuating silver prices left many mine owners and managers feeling uncertain, and they again stressed that the only way for the mines to stay in business was for the miners to take a pay cut. With no ore coming from the mines, the smelters closed too, including the Smith-Moffat–owned Bi-Metallic, which closed on July 1.[27]

Eben Smith's daughter Cora wrote her friend Fanny Watson that with the Leadville mines closed, there were "thousands of idle men walking the streets with nothing to do and no where to go." Fanny Watson wrote to her fiancée in Ireland that the great stand-by and sole support of many of the mining families was "slowly being crushed out of existence" and that the ruins of Pompeii did not look as gloomy as a deserted mining camp. She went on to say that her father, Joseph Watson, was ruined but that, fortunately, Eben Smith was also interested in gold, which cushioned the blow. Horace Tabor, Eben Smith's former employer and sometime business associate, was perhaps the most famous victim of the silver crash in Colorado, with his once vast fortune soon wiped out.[28]

In response, the Colorado State Silver League held a convention at Coliseum Hall in Denver on July 11 and 12, 1893. Thirty-nine counties sent delegates, which included Governor Davis H. Waite. Eben Smith attended as one of the twenty-four delegates from Leadville. On the first day of the convention, the "thunderous cheers" from delegates "were heard for blocks in every direction." The delegates chose Charles S. Thomas as chairman of the convention, and he called the meeting to order. George G. Merrick opened the convention with a speech, saying that when the "pioneers" came to the West, "they had an implied contract that every ounce of silver they dug from the mountains would be coined at the rates then established by law" and that the contract had now "been violated." In the afternoon session that day, Thomas said that silver was the "money of God." Although Smith said nothing during the convention, his fellow delegates chose him as a vice president during the July 11 afternoon session.[29]

At the July 12 session, the delegates unanimously adopted an address to the people of the United States. They reported that 15,000 miners were idle in Colorado, and 4,000 more would be out of work if the smelters shut down. If the panic continued, they argued, railroads, stone quarries, and the lead industries would soon fall. As far as the delegates were concerned, silver mining was not only the source of their own livelihoods but the major support for all other industries in the country. They went on to condemn both the US House of Representatives and the US Senate for repealing the Sherman Act out of a sense of panic. The delegates also appealed to Colorado's eastern creditors, warning that if silver collapsed, Colorado would be unable to pay its debts. In the most shocking section of the address, they also appealed to the South. They argued that Colorado's Republican senators had helped save the South from the "humiliation" and "horrors" of being subjected "to the electoral will of your former slaves" and that "you can save us now." The *Rocky Mountain News* reported that following the reading of the address, thousands of men leapt to their feet and "thundered a mighty 'Aye' in approval." The members also chose delegates to similar conventions in St. Louis, Chicago, and St. Paul, although Smith was not among them. His love for silver was almost supreme, but he remained quiet at the convention and returned to Leadville almost as quietly.[30]

Despite the statement, the federal government took no action, and after the initial shock, the silver market began to recover in August. The smelters reopened mid-month, followed by a number of mines in late August and early September. Smith, though, refused to reopen his mines until reaching an agreement with his workers on wages. At a meeting with the workers on September 2, 1893, Smith and his fellow mine managers agreed to resume mining with a sliding wage scale based on the price of silver. Two weeks later, at another meeting attended by Smith, the two sides agreed on the scale: if the average price of silver was less than 83½ cents per ounce for a month, the miners would be paid $2.50 per day. If the average price was above that, they would be paid $3.00. In early November, the Smith-Moffat mines reopened.[31]

Leadville was good to Eben Smith. It allowed him to rebuild his fortune, and his selection as one of the delegates to the Silver Convention proved that his reputation as a mining giant was still intact. The list of his contributions to the mining industry grew as he illustrated what a few pumps in the district's wet mines could do. He also played a small part in what has become the legendary story of Molly Brown by hiring her husband as superintendent of the Ibex Company. But by the early 1890s, the Smith-Moffat mining interests had spread to the new mining district of Cripple Creek, and it was there that he now turned a portion of his attention.

4

"I Handle All the Mining Business"

Prior to the 1890s, only one man, Robert Womack, had prospected in the Cripple Creek area. Womack was an "unkempt and frequently hard-drinking ranchhand," but he was convinced that there was gold in the hills around Cripple Creek, and he prospected in the area for twelve years before finally finding gold in Poverty Gulch in 1891. Although he became the "father of Cripple Creek," Womack was too poor to develop his claim and sold it for $500. Within a short time of Womack's discovery, miners from all over Colorado began rushing into Cripple Creek and snapping up the particularly desirable mines located on Battle Mountain. Among the new arrivals were Winfield Scott Stratton, James Burns, Warren Woods and his sons Frank and Harry, and members of the Guggenheim family. Production in the district rose from around $500,000 worth of gold in 1892 to $19.5 million in 1899.[1]

The Cripple Creek boom was the last big mining boom of the nineteenth century for Colorado and the last in which Eben Smith would be involved. Joining the rush in early 1892, he and David Moffat paid $65,000 for the Victor Mine, which was located on the northern slope of Bull Hill. The mines

DOI: 10.5876/9781646421794.c004

FIGURE 4.1. The main buildings of the Victor Mine, 1896. Annual Report of the Victor Gold Mining Company, 1896.

located on Bull Hill were not as rich as those on Battle Mountain, but they were still profitable. Smith, who was general manager, was still living in Leadville and Aspen at the time. C. H. Aldrich was in charge of the mine on the ground, and by May 4, he was able to report to Smith "the Victor is looking fine." The town of Victor eventually grew up around the mine. Pleased with their success at the Victor, Smith and Moffat decided the time was right to buy more mining properties in the Cripple Creek district.[2]

On June 15, 1892, Eben Smith and L. D. Roudebush organized the Anaconda Gold Mining Company for the sole purpose of purchasing and controlling all of the mines on Gold Hill owned by the Anaconda, Work, Coronado, and Lone Start companies. Moffat served as president of the new company, while Smith served in the familiar position of general manager. By 1895, the company owned 150 acres of land, and the value of the ore taken out of the mine ran $1 to $50 a ton in gold. A newly constructed mule tramway capable of hauling ten to fifteen cars per trip brought the ore out of the mine. In addition, there was a powerhouse, an engine, boilers, compressors, an air drill, and a blacksmith shop—evidence of just how profitable it could be for a mining company to be bought out by Smith and Moffat. They capitalized the company at $5 million, with 1 million shares worth $5 each. Smith and Moffat controlled 400,000 shares between them.[3]

As 1894 rolled around, Smith was optimistic regarding his Cripple Creek mines. One of the major problems facing all of the Cripple Creek–area mines, including the Victor and Anaconda, was the difficulty of shipping ore to the smelters. No railroads came to the district, forcing miners to haul ore to the smelters by wagon, which was very expensive. Smith and Moffat decided to solve the problem by building a narrow-gauge railroad that would connect with the Denver and Rio Grande line in Florence. The railroad covered a total of 40 miles between Florence and Cripple Creek and cost $850,000 to build. Smith and Moffat bonded the railroad for $1 million, and by 1895 it was earning $60,000 a year on the bonds, with an additional 15 percent per annum on the stock. The Florence and Cripple Creek Railroad finally made it into the Cripple Creek district on May 26, 1894, but by then the area was in the grip of a major labor strike.[4]

Cripple Creek's first unions, the short-lived Fremont Carpenters Number 506 and the Miners' and Prospectors' Protective Association, formed in 1892. The Cripple Creek Carpenters Number 298 operated from 1893 to 1894, and miners in what would become the town of Anaconda formed a union in 1893. Shortly after the founding of the Western Federation of Miners (WFM) in 1893, miners at Altman asked Alexander McIntosh, the Colorado WFM organizer, to help them organize a union. He founded the Free Coinage Miners' Local Number 19 on April 20, 1893, and in a short time, more than 300 men had joined. Within two months, miners organized locals in Cripple Creek, Victor, and Anaconda.[5]

As a Republican, Smith most likely shared in the party philosophy that it was acceptable for workers to unionize but unacceptable for them to strike. Smith and Chaffee dealt with one of the state's first strikes at their Bobtail Mine in 1863, when miners protested management paying them in inflated paper money instead of gold. Smith and Chaffee quickly settled the strike in their favor. The Bobtail was involved in yet another strike in 1873, when miners protested the reduction of their wages from $3.00 a day to between $2.50 and $2.70. When the Cornish miners at the Tam O'Shanter in Ashcroft struck against the non-Cornish foreman, Smith was there as well.[6]

Shortly after the WFM became involved with union activity in Cripple Creek, wages and hours became burning issues. Wages ranged from $2.50 to $3.00 a day, and shifts ranged from eight to ten hours at the Cripple Creek mines. A number of mine owners, described by historian Elizabeth Jameson as more "working-class"–type owners such as Stratton, Burns, and Doyle, ran nine-hour shifts at their mines. The Smith-Moffat mines ran ten-hour shifts and "were among the most intransigent." According to a history of the strike printed in the *Great Divide* magazine in July 1894, the conflict arose as managers at the mines running nine-hour shifts urged their fellow managers who ran only eight-hour shifts to adopt the nine-hour schedule. The nine- and ten-hour mines immediately ran into trouble with the union.[7]

On April 17, three days before the formation of the Free Coinage union, H. E. Locke, superintendent of the Isabella Mine (located next to the Victor property), announced that he was increasing the workday from eight to ten hours, with one hour for lunch. The miners refused to work the new hours, and none reported to work the following Monday. Locke, after receiving orders from Isabella owner J. J. Hagerman, relented and went back to an eight-hour day. The conflict gave mine owners a sense of the trouble the union could cause them, and they met to devise a uniform shift schedule while the WFM began working toward a $3 salary for an eight-hour day agreement at all of the Cripple Creek district mines.[8]

The union made the first move, asking Frank T. Sanders, manager of the Burns Mine, to adopt an eight-hour schedule in December 1893. Sanders refused, and on December 26, a group of more than 200 miners called out the workforce at the Burns. The interested mine owners retaliated by planning to announce on January 20, 1894, that they would enforce a ten-hour day beginning February 1. As each day passed, the tension in the camp mounted, especially when one anxious mine manager, A. D. Jones of the Pharmacist, put the notice about the new hours up on January 17, three days early. The tension reached a breaking point on January 20 when a group of miners ran Locke and another man out of the camp. Still hopeful, miners met two days before the February 1 deadline and agreed to compromise if the mines that worked eight-hour shifts agreed not to extend those hours. Mine owners did not respond, and on February 2, the miners gave management ten days to meet their $3 pay for an eight-hour shift demand. Eben Smith, unwilling to meet the demand, sent a note to Milo Hoskins, who worked at the Victor, advising him that "in case any of our men go out . . . pay them off and [do] not give them any more work." Seeking to undercut the miners, the owners and managers announced that they would close the mines on February 7. That day, the union struck all mines running more than an eight-hour shift.[9]

The strike was relatively small, as fewer than half of the working mines in the Cripple Creek district and less than one-fourth of the 800 miners who had joined the WFM were actually involved. Thomas M. Patterson, writing in a history of the strike for the *Great Divide*, said that three-fourths of the mines in the district ran eight-hour shifts and that only one-fifth of the miners were on strike. Management at a few of the mines, such as the Kismet and Santa Rita, quickly settled with the striking miners. James Burns, one of the owners of the Portland Mine, negotiated a deal with the union in which he agreed to pay miners who worked nine hours $3.25 per day while those who worked eight hours would earn $3. On March 19, Winfield Scott Stratton's Independence adopted the eight-hour day. Other mining companies, though, including Smith-Moffat, were not the least bit interested in settling.[10]

On March 14, a number of Cripple Creek district mine owners—including Smith and Moffat, Hagerman, William Lennox, and Edward De La Vergne—went to court to get an injunction against the striking miners. A district judge ordered the strikers not to interfere in any way with the operation of the mines. With the ruling in hand, the Victor and Anaconda, along with De La Vergne's Summit and Raven, tried to open. No more than five men reported to work at each mine. Fear of the unknown, coupled with the unsuccessful attempt to reopen, prompted the superintendent of the Victor to ask Cripple Creek Sheriff M. F. Bowers to send deputies to protect the mine. Bowers deputized six men, but a group of striking miners intercepted and disarmed them near Altman.[11]

That same night, Sheriff Bowers asked Governor Davis H. Waite to send troops to Cripple Creek. Waite sent three companies, and Bowers boosted his own force by swearing in an additional fifty deputies, most of them ex-police officers from Denver. The troops, under the command of Brigadier General E. J. Brooks, arrived in Cripple Creek on March 18, but Waite recalled them on March 20 after further reports from the district convinced him that they were not needed. On April 1 the striking miners met with mine owners and managers in an effort to reach an agreement. The owners, led by J. J. Hagerman, offered $2.75 for an eight-hour shift with a twenty-minute lunch break, but the miners unanimously rejected the offer. By then, the strike affected only seven mines, and some local businesses were urging the holdouts—including Smith and Moffat—to give in to the miners' demands, not out of any particular sympathy with the miners but because the strike was beginning to hurt their businesses. Smith and Moffat ignored their pleas.[12]

Near the end of April, Smith received a letter from M. Parker, a business associate who was helping him house the scabs who were trying to work the mines. After asking Smith to send him a check to cover some of his expenses, Parker went into the details of a plan he had to break up the union. He wanted to send a small group of men to each of the union precincts, such as Victor, Altman, and Cripple Creek, and cause trouble, keeping the unions busy in their own territory and preventing them from joining into one large body. Parker told Smith that the union was only truly troublesome when all precincts joined together, and he saw his plan as a way to keep the strikers from gaining further power. Smith never instructed Parker to put his plan into action.[13]

In early May, desperate to resume production, some mine owners offered to buy arms for El Paso County and pay for additional deputies to reopen the seven closed mines in the Cripple Creek district. Smith and Hagerman bought 100 rifles and 10,000 rounds of ammunition during the "war at Cripple Creek," but Smith refused to let them be used and kept them locked in a gun store for the duration of the strike. In large part, he did this to protect both sides, as Smith wrote in a letter to Charles A. Keith at the Anaconda that any guns sent there would be "very likely to fall into the hands of the strikers as did the 125 sent down from Leadville." As late as July 1894, with the strike all but over, Smith still refused to allow Hagerman to use the guns on raids in case they were needed somewhere else.[14]

By mid-May, the situation was incredibly tense. Sheriff Bowers deputized almost 1,200 men and sent nearly 200 of them to occupy the closed mines. The striking miners, fearful that Bowers was getting ready to attack them, built a "fort" on Bull Hill that commanded nearly all of the mines on that hill. The fort caused even more fear among the mine owners, and on May 17, they asked the El Paso County Board of Commissioners to send forces to protect the mines. On May 24, 150 to 200 of Bowers's deputies started marching toward the town of Victor, and the miners on Bull Hill marched down to meet them. The miners gathered near the mines on Battle Mountain and seized the Strong Mine. Once in control of the property, they ordered the superintendent, foreman, and engineer to come out; fearful, the three went down into the shaft. The striking miners then set off an explosion that destroyed the shaft house, trapping the three men inside the mine and frightening the deputies into retreating along the Florence and Cripple Creek tracks. The next day, as 300 of the strikers set out to attack the deputies, an advance party ran into an outpost set up by the deputies. Both sides fired shots, killing a deputy named Rabideau and a miner named Herman Crawley. The deputies arrested six striking miners and took them to Colorado Springs. In retaliation, the striking miners rescued the three men trapped inside the Strong and held them hostage.[15]

As the violence increased, so did the pressure for a settlement, and on May 26, Governor Waite asked the strikers to lay down their arms. He also declared that Bowers's deputies were illegal, based on the state attorney general's opinion that it was illegal for Bowers to employ deputies from outside his own county. That same day, a committee from Colorado Springs went to Bull Hill at Hagerman's request and successfully arranged to trade the six jailed strikers for the three hostages. The next big move in the strike came on May 30 when the striking miners met with Governor Waite and gave him the power to negotiate for them. The pressure for a settlement was intense, but the sympathy of many in the mining business was still with Smith and his fellow mine owners. In early March, Bela Buell, an old friend of Smith's from Central City, wrote to him "I congratulate you on the stand you have taken so far" and said he wanted to see the strikers forced to go east to work on farms for four or five years so that "perhaps then they will appreciate a good thing." Simeon Jordan, one of Smith's former stepsons, wrote to him "we sympathize with you people and hope you'll kill the last one of those rebels as I term them."[16]

Smith's visits to his Cripple Creek mines were less frequent than usual during the strike, but on May 28, he went there specifically to survey the situation. He hated what he saw. The next day he wrote to J. J. Hagerman that the only thing to do was get writs for those who had broken the law and then recruit a sufficient force to serve them since the sheriff was "as usual sitting around with his thumb in his mouth." The following day, Smith wrote to Hagerman that the Rouse and Fremont miners had gone out on strike and that the Fremont miners were likely to try to reach Cripple Creek on the Florence and Cripple Creek Railroad. Smith ordered a man at the railroad named Johnson to delay them until he could send out an engine to destroy one of the bridges if this did come to pass, at the very least delaying, if not preventing, them from getting to Cripple Creek altogether. Fortunately, the end of the strike saved Smith from having to destroy one of his expensive bridges.[17]

Although pressure for a settlement was intense by the beginning of June, many felt the strike was likely to continue, as fellow mine owner John F. Campion telegrammed Smith that the Knights of Labor had pledged to support the striking Cripple Creek miners. However, Governor Waite and union organizer John Calderwood met with Hagerman the next day, and he reluctantly agreed to the miners' demands, although his fellow mine owners convinced him not to sign the agreement. This last show of force by the mine owners achieved nothing, and two days later Hagerman, this time accompanied by Moffat, met with Waite and again agreed to the $3 pay for an eight-hour day deal. Both men signed the second agreement. Even then, some were unhappy with the outcome. Franklin Ballou, vice president and general manager of Smith-Moffat's Bi-Metallic Smelter, wrote to Smith on June 5 that the settlement of the strike would have been better "for the future of mining in this state if those scamps had been brought to terms by force." Joseph Watson, who according to his grandson Louis O'Kane would never even own a gun, told Smith he "wanted to see about 50 of them knocked into eternity."[18]

Moffat and Hagerman may have signed the deal effectively ending the strike, but that did not end the trouble. Bowers and his illegal deputies were still on the prowl, and on June 4, 1894, Waite had to call out the state militia to try to stop them from marching on Bull Hill. Weather kept the militia from reaching the Cripple Creek district until June 7, but by then Bowers and his deputies had already exchanged shots with some of the strikers. Only when General Brooks threatened to fire on the deputies did they finally retreat from their positions. Rather than disband, though, they continued on to Altman and then to Cripple Creek, where they arrested several people and took charge of the Independence Mine. In an effort to try to end the trouble, prominent mine owners met with militia officers and county officials on June 10. Only after that meeting did the deputies disband, and control of the mines returned to their proper owners after the military guaranteed protection for mines and mine owners.[19]

In late June, William Trevorrow, who worked for Smith in Cripple Creek, reported that his wife had received a number of disturbing letters. Smith wrote back, saying "there are a great many mortifying things occurring to all of us at present." He added that "some irresponsible tramp" who did not know what else to do and had heard that Moffat "was the cause of all this trouble" had broken out all the windows in the First National Bank of Denver building on July 8. Smith told Trevorrow that if it was necessary to keep a standing army at the mines in order to keep them open, it was better that they stay closed. Bitter over the final settlement, Smith saw the chance to elect a "decent governor" in the next election as the mining industry's only hope. Authorities filed the first criminal charges resulting from the strike in June 1894. Attorney J. E. Rockwell wrote to Smith on August 1 that no one would be "persecuted" but that everyone who had committed a crime would be "prosecuted to the limit." In the end, those who were convicted of any crimes were either pardoned or had their convictions overturned.[20]

Smith and Moffat's harsh response to the striking miners in Cripple Creek, as well as during the strike that would rock Leadville two years later, was in stark contrast to the way Moffat treated workers on the Moffat road in the early 1900s. Edward T. Bollinger, who did a great deal of research on the construction of the Moffat road, wrote that Moffat "set a noteworthy example of labor-manager relations with his rank and file men whom he trusted implicitly event [*sic*] to the writing of contracts." Moffat also frequently visited with workers in the crew compartment of the locomotives on the train and insisted that they be fed first, even when investors were traveling on the train. This difference in his behavior is probably best explained by the fact that for Moffat, mining was a means to an end, with the end being construction of his railroad. Moffat's biographer, Steven F. Mehls, writes that the harshest response to the striking miners came from owners who had not actually worked as miners at some point in their careers. This would again explain Moffat's response but not Smith's, who had indeed worked as a miner. Smith frequently told those who wrote asking him for money that hard work would solve all of their problems. Perhaps having risen to become a manager and mine owner himself and having made and lost a fortune, he had little sympathy for those who to him seemed unwilling to work hard and follow the same path.[21]

The strike was over when the Victor, closed since the beginning of February, reopened in August 1894, but there was still trouble at the mine. In September, Smith heard reports that miners were stealing ore from the Smith-Moffat mines in Cripple Creek. Smith instructed Superintendent Charles Keith at the Anaconda to build a change house at the Victor and require that all men change their clothes at the end of their shift, saying "the honest man will not object and the thieves we do not care to employ." Despite Smith's faith, this plan did not work, and in April 1895, he hired Pinkerton's Detective Agency to find out who was stealing his ore. The Pinkerton operative, known only as W.B.S., wrote to Smith on April 27 that a "third party claims there is a systematic scheme on foot to rob the Victor Mine" and that those running the scheme stole eleven sacks of gold ore from the mine in February alone. The plot was actually fairly clever. A miner would lease a supposedly worthless claim in the area and then present the stolen Victor ore as having come from that claim. Rumor had it that several people were involved in the plot, including the assayers in the camp, who would knowingly buy the stolen ore. W.B.S. started passing himself off as a mining expert hired by Smith and by May 8 had focused his attention on a miner named Andy Malloy. The investigation came to a sudden halt when Smith, in a letter dated May 16, ordered him to drop the investigation since he had decided to get a new, trustworthy crew at the mine.[22]

Smith dismissed a number of men, including the superintendent and foremen, but the problem persisted. On May 23, a new Pinkerton's operative, G.H.S., was on his way to the Victor. He spent several days posing as a miner looking for work in the Cripple Creek district and was "hired" to work at the Victor on May 31. As part of his investigation, he learned that there were still a number of sacks of ore hidden in the mine and that it was easy for someone to steal the ore because the men were not required to change their clothes when they took their lunch breaks. G.H.S. learned that a man named Jack Smith, who had recently been killed, took part in the ore theft because "he hated old Moffat's guts." The investigation also revealed that several thousand dollars' worth of ore had been stolen. On June 5, before G.H.S. could get any deeper into his investigation, someone started a rumor that he was not really a miner, forcing Pinkerton's to recall him the next day. The investigation went no further, though James McParland, assistant superintendent of Pinkerton's Western Division, told Smith he had another operative available whose cover was foolproof.[23]

During the Cripple Creek strike, Smith was content to let Moffat be the public face of their mining interests while he worked behind the scenes from his new office in Rooms Five and Six of the fashionable Tabor Block in Denver to expand their interests. On February 1, 1894, he incorporated the Bon Air Mining Company in Leadville and assumed its presidency. Of the 400,000 shares of stock available in the company, originally worth $1 apiece, Smith controlled 128,999 shares. Thurlow Weed Barnes later wrote to Smith from New York that he was the "sort of straightforward and competent man that I had always hoped would come into control of this property." Three months after incorporating the new mining company, Smith helped incorporate the Gold Extraction Company of Colorado, and the directors elected him president and hired Philip Argall, an expert smelter, to operate it. This company set out to erect a 100-ton-per-day cyanide reduction works at the intersection of the Florence and Cripple Creek and Denver and Rio Grande Railroads in Florence. This location was also near the Atchison, Topeka and Santa Fe Railroad, enabling the mill to receive ore from Cripple Creek, all other parts of Colorado, Arizona, New Mexico, Old Mexico, "and every other producing camp in the Rocky Mountain Range." It mainly handled Smith-Moffat ore; according to the incorporating documents, the company had already received a favorable contract from the Smith-Moffat–owned Florence and Cripple Creek Railroad, and the directors estimated that the first-year profit would be $124,000. Just one month later, on May 16, 1894, Eben Smith, David Moffat, and Charles J. Hughes incorporated the Resurrection Gold Mining Company in Leadville. Smith, Moffat, and Smith's son-in-law Charles Carnahan served as directors of the new mine, and Carnahan assumed the duties of superintendent. The new company issued 100,000 shares of stock at $5 each, all of which Moffat or the Smith family controlled.[24]

By the early 1890s, Smith was again heavily involved with stocks, both mining and otherwise. At a meeting of the board of directors of the Victor Gold Mining Company on June 16, 1894, he successfully offered a resolution to list the Victor stock on the New York Stock Exchange. After the exchange accepted the company's application, the stock—handled by H. L. Horton and Company—sold quickly, especially in France. This created a huge new pool of capital that Smith and Moffat could reinvest in the mine, in addition to making the owners of the mine very rich. H. L. Horton and Company did such a good job that in December, Smith told them he was opposed to selling any more shares, writing to Moffat that the value of the mine had doubled and he wanted to "retain" the property, "but if you think it advisable to make a new arrangement with those people [H. L. Horton] or with others to sell the balance of the stock, I will try in some way to accomodate [*sic*] my views to yours."[25]

Moffat did such a good job of convincing Smith that by January 1895, he and Emily were in New York preparing to leave for London so he could assist in selling the stock there. New York was one of the Smiths' favorite destinations, and while there in January, Eben found himself "studying there [*sic*] ways and manners." Smith had started wearing dentures in 1893, and one of the ways and manners that most impressed him was the fact that many New Yorkers found it fashionable to have two sets of false teeth. He only had one set with him, so he asked his new secretary, Robert Reid, to get his second set of false teeth from his house and send them to him. Smith was pleased to find himself among the fashionable set in New York, but for reasons unknown the Smiths canceled the trip and returned to Denver.[26]

Taking over for Smith, David Moffat arrived in New York in May 1895 to do his part to help sell the Victor stock. Although Moffat may have been the acknowledged financial brain in the operation, Smith was not completely uninvolved in that side of the business, and Moffat trusted his advice. Shortly after Moffat arrived in New York, he wrote to Smith that he did not know what the price of the stock would be but that it would not "be lower than what you told me I could go before I left Denver." Stock sales in France continued through July, and Horton himself wrote to Smith, saying "you cannot imagine how many d—d questions the French people ask. Talk about Yankee inquisitiveness, but it cannot be compared with the Frenchman when it comes to trading in mines."[27]

Selling stock was a great way to make money for further development of the mine and increasing personal wealth, but Smith was also very interested in buying other stocks in hopes of adding to his fortune. In 1893, he purchased $10,000 worth of stock in the Battle Mountain Gold Mining Company, selling it the next year at a profit of a little over $1,100. His broker demanded a healthy commission from the sale, but Smith thought the amount was too high and refused to pay it, much to the dismay of his associate William Lennox. He wrote to Smith that if in his future business dealings he would "learn to be a little more honorable, and make good your verbal promises you would in all probability avoid a great deal of annoyance and trouble." Smith usually thought the commissions he had to pay were too high. In 1902, when another associate informed him that he had agreed to a 10 percent commission, Eben answered that if it were true, he would have "made a roar that you could have heard clear across the country."[28]

Incorporating two mining companies, selling stock in a third, and dealing with a labor strike in addition to managing his existing mines would probably have been enough in one year for most men, but not for Eben Smith. On September 1, 1894, Smith and Moffat bought 5,000 shares of the Baka-Contact Gold Mining Company in Cripple Creek for $5,000. As part of their purchase, they agreed to erect a ten-stamp mill and build a wagon road to the mine, once again demonstrating how beneficial their ownership was for a mining property. Still searching for even more mines, in late 1894 Smith initiated a major change in Smith-Moffat policy by purchasing a mine outside Colorado.[29]

Several of the many offers of mining properties that regularly crossed Smith's desk concerned properties in Mexico, but Smith always rejected them. In August 1894, Smith responded to an earlier letter from E. C. Foster, declining a mine in Zacatecas, Mexico, telling Foster that the mine was "too far away from Colorado to interest me." One month later Smith and Moffat received a letter from John S. Crawford offering them a silver mine in Mexico. Smith answered "I handle all the mining business of our concern," adding that he and Moffat generally "refused to go outside Colorado . . . to invest in mines." In addition, the decline of silver had led them to decide "not to touch any silver mines until something more has transpired in regard to the future of silver." When it came to gold mines, however, Smith was more than willing to venture outside the United States. In November of the same year, Smith and Moffat paid $30,000 for the Santa Elena Gold Mines in Sonora, Mexico, incorporated a new company, and issued 500,000 shares of stock.[30]

This additional activity required a great deal of Eben Smith's time and attention, but not so much that he neglected his existing mines. The load was somewhat lightened due to the fact that the Victor and Anaconda were not operational during much of 1894, but there was still more than enough to keep him occupied. In August, he wrote to Lee Wood at the Smith-Moffat mine in Rico, Colorado, that he was sorry he had been unable to visit the camp recently, which did not look "very well for a general manager," but "as the general manager receives no salary and has not more than a million shares of the stock he may be excused." In spite of his absence from the camps, Smith still wanted his foremen to keep him posted on how the mines were doing, and he expected full details. For example, the letters he received in the summer of 1894 from Ronald Morrison, who was in charge of the Smith-Moffat Golden Eagle Mine in the Cripple Creek district, went into incredible detail about the depths of shafts, surface water, and vein size. Smith was concerned about every detail, no matter how small.[31]

The mine-buying spree continued into 1895. In January, Smith leased all of George E. Collins's mining claims in Chaffee County, twenty-one in all, for a total of $20,000. In February, Smith and Moffat signed a contract to buy 5,100 shares of stock in the Antlers-Park Regent Leasing Company in Creede, Colorado, for $10,000. Even with the injection of $10,000 from Smith and Moffat, the company soon ran out of money to continue operations, and at a July stockholder meeting the board of directors agreed to increase the capital stock from $50,000 to $100,000. Smith and Moffat agreed to buy half of the additional $50,000 in stock, but only after the other half sold. The increase in capital stock seemed to work because in October the president of Antlers-Park, E. Barnett, wrote to Robert Reid that the ore on the sixth level of the mine was running from $16 to $26 worth of gold per ton and that they were fast "approaching the day when we shall get rich dividends from our property."[32]

Smith and Moffat's interest in Antlers-Park did not make everyone happy. O. P. Poole, who worked for Antlers-Park and would be associated with Smith on and off for a number of years afterward, wrote to a friend that the two mining men were not that interested in the details of running a mine. Instead, he wrote, they were "people who would rather have or think they had some one else to look after such details as their time is to [*sic*] much taken up with other larger affairs." There were others, though, who felt just the opposite. Some were so pleased that they even offered to sell Smith and Moffat additional property. J. S. Jennings wrote to Smith in June 1895 offering to sell him his claims below the Antlers Park region because Jennings was "told by all that know you that you are a fair man to do business with."[33]

Jennings was not the only one who thought that way, and some were in desperate need of help from the fair man. By the end of January 1895, United Mines, also in the Cripple Creek district, faced serious trouble. E. M. Ray, superintendent of the mine, received a letter from J. A. Dean on January 23 about the company's inability to pay the men working there. Dean wrote that he hoped Ray would be able to raise the money to pay the men who had worked at the mine and let them get out before trouble started. At the time, Dean wrote, they were "quiet and orderly, but I can see that a feeling of uneasiness and dissatisfaction prevails among them."[34]

In an effort to save the company, the directors reincorporated it as the United Leasing Company and made Eben Smith a director as well as the company's manager. Smith tried to make a success of the United, and in his mind, keeping expenses down at the mine was an important objective. On June 18, 1895, he wrote to E. M. Ray to "please keep me posted from time to time how the level looks and also how the mine looks generally." He went on to advise Ray to reduce the number of pump men at the mine from four to three and asked what the mechanics were doing to justify their pay of $4.00 and $4.50 per day. He told Ray, "I have no disposition to be critical in making these inquiries, but a property like this with no income should be managed with the greatest care in regard to expenses and I trust you will bear this matter in mind." The following November, Smith became president of United Leasing, but his increasing number of titles and his work at the mine were no more successful at saving it than the efforts of those who came before him. On January 13, 1896, W. H. Bryant sent a letter to Smith calling for a stockholders meeting, at which it was resolved to suspend work on the property and forfeit the lease. The failure of the mine led to a series of lawsuits and liens against the company, such as one filed on January 16 by Thomas Armstrong for $6,984 on behalf of forty-nine men who worked there. Then, on February 6, he filed a Writ of Attachment for the same amount on behalf of the same workers.[35]

The dissatisfaction did not end with Armstrong. On January 17, Robert Reid received a letter from W. C. Bowers, a stockholder in United Leasing, who complained that at the time he purchased more stock in the company, it had already "practically failed" and he felt cheated. On February 10, 1896, David Moffat, who was less involved in United Leasing than Smith, received a letter from the firm of Barnie and Rue, asking when they could expect a settlement from the mine. In mid-February Armstrong, pursuing the claims of the forty-nine workers against the company, wrote to Reid offering to settle the claims for 95 cents on the dollar, which amounted to $6,984. He did not like making the offer but wrote that he did so because "there is terable [*sic*] hard times here and some families are in bad circumstances." A little over a week later, Armstrong wrote to Moffat in an effort to find out if the miners' claims against the company were worth anything. Men were selling their time checks for 50 cents on the dollar, and if they were worthless, Armstrong was going to tell them to "get out of here as best they can." After this letter, Smith and Moffat heard no more on the matter, and the United Leasing Company went down as one of the few but perhaps the biggest failures of Eben Smith's mining career.[36]

Even as he struggled to save United Leasing, Smith continued to take charge of other mining properties during 1895. Early that year, Smith and Moffat signed a contract to purchase all the stock of their former ally in the 1894 strike, the Raven Gold Mining Company, which held the Raven, Princess R, Snowy Range, Maid of Erin, and Gregory Lodes, all in Cripple Creek. In the meantime, Joel Parker Whitney decided that he wanted to buy all the property and rights, and on October 22, 1895, he signed a contract with Smith and Moffat that allowed him to do so. They sold out to Whitney for $455,000 plus stock in the company, making a nice profit without having to invest a dime in improvements to the Raven property.[37]

As usual, unsolicited offers of mining properties arrived at Smith's office on a daily basis. In May 1895, he turned down an offer of a mine in Utah from Frank Knox, writing that his and Moffat's interests in Colorado were "so very large and require all of my time and attention to keep track of them." In August, T. H. Norton offered Smith and Moffat the Plomo Mine if they were "not already loaded up with mines and mills." The mine must have been of some interest to Smith because he did not immediately reject the offer. He disagreed with Norton over some of the costs Norton quoted, particularly that it would cost $3 to $4 per ton to reduce the ore to bullion. Norton wrote back that even if the costs were higher, the mine would still be a paying investment and a few weeks later sent Smith a 500-pound sack of ore on which he could run sample tests. By then Smith had either tired of the exchange or decided that the mine was not a good investment, as Norton wrote to him in mid-November that he had not heard the results of the tests. Smith penciled a note on Norton's letter that "the tests of ore were sent to Norton," and the matter was dropped.[38]

Eben Smith made one other major business move in 1895. His son Frank sold his share in Reynolds and Smith Ore Hauling and Heavy Teaming for a reported $2 million that year and moved his family to Denver. While Smith looked for a new business he and his son could get involved with, he was growing increasingly frustrated with what he thought were the high prices mining supply companies charged him. An ideal solution to his predicament presented itself, and on April 23, 1895, Eben, Frank, Robert J. and John S. Cary, and John Y. Oliver founded the Denver Mine and Smelter Supply Company, buying out the Kennedy Pierce Machinery Company and using its stock to start the new business. Eben was president and Frank served as treasurer. By 1900, the company was so successful that it had branch offices in Salt Lake City, El Paso, and Mexico City.[39]

Two months after incorporating the company, Smith wrote to W. J. Chalmers that he expected to make a "swimming card" of the new business and would spare no pains, "either in time or money, to make it the largest machinery concern in this state." Smith felt assured of success because his "acquaintance with the mining people of the West, is probably more general than [that of] any other man in this section." Left unsaid was the fact that the company would also supply all of his and Moffat's mines at a "fair and reasonable rate." With the creation of this company, Smith gained complete control of his mining business. He and Moffat owned the mines, the railroad that shipped the ore, and the smelters that handled the ores; now he owned the store that would supply both the mines and the smelters. No matter what, he was certain to make money.[40]

For the most part, the Smith-Moffat organization ran like a well-oiled machine. Far from an absentee landlord, Smith was intimately and intensely involved with every aspect of his business. Nearly every day he received detailed reports from the superintendents at his mines, and he became upset if these reports did not arrive regularly. As far as he was concerned, no matter was too trivial for the reports. The letters Smith received from Norris H. Cone of the Victor in the 1890s contain a wide variety of information and are good examples of what he expected to read. Cone started working for Smith in Leadville in the early 1890s, moved to the Victor in 1894, and became superintendent in 1895. Between 1895 and 1897, Cone had, among other things, reported to Smith on his efforts to find out why rock from other mines was falling into the Victor shaft, discrepancies in the price of coal charged to various Smith-Moffat mines, and the depths of the shafts at a number of their mines. Cone also made sure to tell Smith that he received his regular instructions, such as being certain to actually ship out the ore taken from the mines. While Moffat might not have understood much in these reports, Smith was able to comprehend them with ease. By the 1890s, he had nearly forty years of mining experience and was often referred to as one of the foremost mining experts in Colorado.[41]

Smith did not simply read these daily reports and file them away. He usually responded with detailed instructions of his own, in addition to visiting and going down into the mines about once a month. If the case warranted it, he might conduct an "interview" over the telephone with the superintendents or, if the problem was severe enough, make an additional visit to the mine. The most common theme in Smith's instructions was to keep costs down. In June 1895, he wrote to Charles Keith at the Anaconda to have a current employee, in this case a man named Hammond, attend to billing and tagging the shipped ore, as "this thing of hiring a man for every little duty we have there to perform is not the thing. The expenses at that mine by adding new men are in my judgement entirely too high."[42]

While Smith's feelings about keeping costs down could be intense, he was most passionate about the smelting of ores. Smith's stunning success in milling in the early part of his career led him to write in 1898, "I believe that I have a right to say that I am pretty thoroughly posted on how a mill should be constructed for producing the best results in savings and economy in running." As such, he could sometimes be unbearable on smelting matters. In February 1895 his secretary, Robert Reid, sent a reminder to Charles Keith at the Anaconda to ship high-grade ore as a separate lot so it would receive a better price than it would have when mixed in with lower-grade ores. Even with these kinds of measures, Smith was in almost constant disagreement with even his own smelters on the value of the shipped ore, and a settlement often had to be made on the split between the smelter's assay and the mine's assay.[43]

With all their new purchases, Smith and Moffat never found a property that replaced the Victor Mine as the showpiece of the Smith-Moffat operations, and as such it received a great deal of attention from Smith. In the 1895 annual report of the Victor Gold Mining Company, Smith was able to write that production at the mine in the year following the 1894 strike was excellent. According to him, the Victor "never looked so well as at the present time," and no mine in the Cripple Creek district had as "true a vein and as great a length of ore chute as the Victor, and none has more promise of paying to a greater depth." The mine was doing so well that the company was paying $20,000 a month in dividends to the stockholders. In January 1896 Smith wrote to Reid that he was "well pleased with the Victor report and only hope they [the miners] will continue to keep up the good work."[44]

In July, Smith was not nearly as pleased, and he wrote to Cone that "you must pay more attention to this property and know more about what is going on." He also instructed him to "push the output [in] the fore part of the month as I once before suggested to you." Cone followed the boss's instructions, and Smith was able to write in the August 1896 annual report that the width of the vein, which had ranged from 1 inch to 10 inches in 1895, was now ranging from 3 inches to 8 feet. The ore was not as high grade, but the quantity more than made up for that fact. Smith wrote that although small veins sometimes ran out, large ones seldom did, and the change at the Victor only strengthened his opinion "of the permanency of the mine."[45]

By late 1897, Smith regarded Cone so highly that he placed him in charge of all the Smith-Moffat Cripple Creek properties, but he did not excuse him from sending regular reports. In early May of that year, Smith stopped in Cripple Creek to inspect the mines, and he was particularly upset with the expenses at the Victor, which were running about $20,000 a month. Smith instructed Cone to cut these costs to $15,000. Even as Smith was preparing to go overseas in 1897, he took time to send a last round of orders. In a letter to Robert Reid, he gave detailed instructions on how to set up a hoist at the Victor and the other Cripple Creek mines, then instructed Reid to abandon the mines if they did not produce good ore.[46]

As busy as Smith kept himself, often working six days a week, he still found time to enjoy a life outside of business. The vast empire Smith-Moffat controlled usually meant that Eben had to combine business and pleasure, but that did not bother him. In fact, he preferred it that way, and throughout most of the 1890s Smith, who was in his sixties, enjoyed his workaholic lifestyle. Eben and Emily moved to Denver in 1893, where they rented a house at 1151 Corona Street while they waited for completion of their new mansion at 951 Logan. The seventeen-room mansion, designed for them by Varian and Sterner, included an "elaborately furnished library," a hall, a dining room, a large drawing room, and a kitchen "furnished with all the most convenient and modern arrangements" on the first floor. The bedrooms and bathrooms were on the second and third floors, all "tastefully decorated." Two staircases

FIGURE 4.2. The Smiths' house on Logan Street in Denver. Denver Public Library, Western History Collection, X-25877.

ran from the top to the bottom of the house. The outside of the house was finished in cream-colored Roman tile with buff brick trimmings. The Smiths moved in on March 26, 1894.[47]

Eben and Emily spent a great deal of time with their grandchildren after their move to Denver. On April 12, 1893, their daughter Cora married Charles Carnahan, who began courting her in 1890. Carnahan's marriage to Cora immediately qualified him for a job at the Smith-Moffat mines, which, in turn, led to his high position when Smith and Moffat incorporated the Resurrection Mine in Leadville. The Carnahans settled into Eben's old house at 120 West Seventh in Leadville and had their first child, Harold Smith Carnahan, on January 27, 1894. Frank Smith and his new wife, Josephine Bonita Hill, daughter of prominent mining man Charles L. Hill and his wife, Josephine, joined them in the house. Frank and Josie's marriage on December 19, 1893, came as a surprise to Denver society, where she was very popular. Their son, Eben LeRoy Smith, was born on October 26, 1894.[48]

In 1895, Frank and his family moved into a house at 1740 High Street in Denver. The Carnahans followed shortly afterward, moving into a house

FIGURE 4.3. Josephine Smith with her son Eben, ca. 1900. Michel Tritt Collection.

FIGURE 4.4. Frank Smith, ca. 1900. Louis O'Kane Collection.

at 1440 High Street, although they maintained a house in Leadville as well. Eben and Emily were thrilled to have their children and especially their grandchildren so close. They bought a high chair, a walking chair, a nursery chair, and a crib from the J. G. Kilpatrick Artistic Furniture and Upholstery Company in Denver and set up a room for the children on the third floor of their Logan Street house. In 1895, a third grandchild joined the family when Doris Leonora Carnahan was born in Leadville on September 15.[49]

Eben even found time to enjoy a good cigar, usually buying them from cigar stands near his office. At home, he enjoyed roasts, veal, and steak for dinners that were prepared by one of the Smiths' maids, "colored Mary" or "fair Emma." For recreation, Smith enjoyed a game of billiards or shooting. If all else failed, there was plenty to read in the Smith house. In August 1895 alone, Smith paid more than $400 for 142 books with which to stock his library. Emily spent a great deal of her time in Denver working with charities, just as she had done in Leadville. She was particularly active in her support of the Old Ladies' Home, Children's Hospital, and the Home for the Aged Poor—hosting dinners for these organizations in the house on Logan Street. Emily was as active in her charitable activities as her husband was in

FIGURE 4.5. Cora Carnahan with Harold and Doris, ca. 1900. Louis O'Kane Collection.

FIGURE 4.6. Charles Carnahan with Doris, ca. 1900. Louis O'Kane Collection.

pursuing mines. After their move to Denver, Eben and Emily also became members of St. Barnabas Episcopal Church on Vine Street.[50]

As the Smiths settled into their Logan Street house, Emily made sure it was pleasant outside as well. She enjoyed buying flowers for the gardens and surrounded the house with one of her favorite plants: lilac bushes. The

Smiths were happy with their new home, located in Denver's "Quality Hill" neighborhood, and with their high position in society. The Smiths lived fairly quietly, though, and were more content visiting with their grandchildren than promoting the state the way David Moffat or others might have done. It was good that Eben was able to enjoy is time outside of work because events were about to take another nasty turn for the Smith-Moffat mining interests.[51]

5

"Our Various Operations"

In early January 1896, Eben Smith and David Moffat arrived in New York City, where they arranged to list the Victor on the Consolidated Stock Exchange. While in New York, Smith continued to receive glowing reports on the Victor from Robert Reid, who was overseeing operations at Smith-Moffat's new headquarters in Room Six of the Equitable Building in Denver. Smith found the reports to be full of "interesting facts." Overall, he was delighted with the Victor and only hoped the workers would keep up the good work. He wrote to Reid that he was enjoying himself immensely in New York and was in no hurry to return to his "sweat-box."[1]

Smith's absence from Colorado did not deter those who wanted to sell him mining properties, and Reid also informed him of the many offers of mining properties that continued to find their way to his desk. In mid-January, Reid sent Smith a telegram informing him that Norris H. Cone had the chance to buy 40,000 shares of Arvilla Tunnel Company stock for Smith and needed to know whether he should do so. Smith wrote back to Reid that he knew "nothing whatever about Arvilla tunnel or the value of the stock

DOI: 10.5876/9781646421794.c005

or where it is located and why in the Devil I should buy it is more than I can comprehend." Cone's persistence in the matter, however, was so persuasive that Smith ended up buying all of the Arvilla stock. Most of the offers that arrived at Smith's office, though, were from total strangers who were unable to plead their cases as effectively. He received so many that in March 1896, he wrote to his friend William Johnson that he wished he could get out of the country and "go some place where I would not be annoyed to death by fellows who want to sell me mines."[2]

Smith may have been in no hurry to return to his sweatbox, but late on February 4, Eben received word that his brother Henry, manager of the Smith-Moffat Leadville properties since 1892, was dying. Unable to leave New York until the seventh, Eben instructed Reid to "see that his pretended wife" did not get away with any of Henry's property. The next day, Smith confirmed his departure time with his secretary and asked him to "notify my foremen everywhere." By the time Eben returned to Colorado, Henry was dead, and it fell to him and their brother Mark to settle the estate. This was somewhat difficult as Henry, according to Eben, had done some very weak things in the last year of his life, particularly where women were concerned. While Eben handled the business of Henry's estate in Colorado, Mark returned to Pennsylvania, where Henry owned a farm, to supervise business there. Henry's estate was finally settled by spring, and under the directions of his will, Eben started sending monthly checks to their sisters Nancy Bassett for $50 and Emily Lyons for $25. Eben not only continued these monthly payments for the rest of his life after the money from Henry's estate ran out but also added a $25 monthly payment to their former sister-in-law, Jennie Smith Gray, who was then married to one of their cousins.[3]

Eben Smith was usually willing to help his family financially, but his increasing fame and wealth led a number of strangers to write to him seeking charity. Smith had a reputation as a demanding person when it came to his mining affairs, going so far as to write to Cone in 1897 that he was "a crank on mining matters." Given this reputation, many might have expected Smith to take a similar stand toward those who were seeking help, but in reality, most people found him an incredibly soft touch who was ready and willing to part with his money for a good story. Mary Johnson, a neighbor of the Smiths on Logan Street, wrote in September 1894 seeking financial help.

Smith agreed to pay her $3,000 for nine lots she owned in Denver "for the purpose of helping you in your distress." In Johnson's case, he did get some property in exchange for his help, but such an occurrence was rare.[4]

In May 1896, William Curnow, a worker at the Bon Air in Leadville, died two days after he was injured at the mine. Smith wrote to a business associate that the company was not responsible for his death since he had no business being in the mine at the time of his injury, but in late June he received a letter from Curnow's widow. She wrote that she had three small children to raise with no means of support and that she had only $50 to pay the $250 cost of the extravagant funeral her brother-in-law had arranged for her husband. Smith wrote to her that he believed the company had done all it was its "duty" to do and that while he was sorry that the handling of her husband's body was not in accordance with her wishes, the company could do nothing about it. In late July, Mrs. Curnow wrote back, telling Smith "I will say there isn't much charity around your heart." That got to Smith, and in early August he instructed Arthur Nichols to send Mrs. Curnow $150.[5]

Smith also did what he could to find jobs for those who wrote to him in search of one, even if he thought they were unfit. E. J. Sweet wrote to Smith in March 1897, seeking a job as an engineer on the Florence and Cripple Creek Railroad. Smith thought Sweet was too old for the work, but he told A. C. Ridgeway, superintendent of the Florence and Cripple Creek, that Sweet's case was the "old story repeated having been an old friend of mine in years gone by and I having shown a disposition to aid him there is no limit to the demands for further aid." Smith could talk tough, but more often than not, his soft side came through. He gave Sweet a less demanding job, with the potential for promotion to conductor.[6]

Shortly before his death, Henry Smith reported that some Leadville miners had threatened to strike unless management increased their pay to $3 per day, which he said would have been fine if the price of silver was higher. The talk of a strike intensified after the Western Federation of Miners (WFM) met in Denver in mid-May and elected former Idaho state senator Edward Boyce as president of the union. Boyce downplayed the strike talk, but on May 23 he, along with leaders of the Knights of Labor and Cloud City Miner's Union Number 33, met with Smith and other mine managers to request the increased pay. They all refused, and when the union leaders again approached Smith and S. W. Mudd, Smith asked them to put the request in writing. They did, but he again refused it. The mines that were paying less than $3 a day were silver mines, and, according to John Campion, they could not afford to pay more due to the low price of silver.[7]

The failed meetings between union leaders and the mine owners created a very tense situation in Leadville, and Smith and his fellow mine owners took steps to protect themselves. On June 12, Smith, John Campion, S. W. Mudd, and A. V. Bohn founded the Colorado Mines Association. In the founding agreement, the men wrote that the "quiet, peaceable and legal operation" of the mining properties that they either owned or controlled was prone to interference from people who had "no legal nor [*sic*] equitable right to do so." The men pledged themselves, "each to each," to "decline to submit to such illegal and inequitable interference, from whatever source it may emanate." The men went on to say that they would shut down the mines if they became too difficult to operate.[8]

The mine owners and managers also intensified their campaign to get rid of Michael Newman, the recently elected Populist sheriff of Lake County, because they doubted his willingness to protect their private property. Smith and Campion initially led the campaign against Newman, but it was other mine owners and managers who eventually hired Pinkerton's detectives to search for a reason that would let them remove him from office. Just two months after Newman took office, deputies arrested him on charges of blackmail, extortion, and oppression. Authorities eventually dropped the charges due to a lack of evidence, but the opposition continued its campaign against the sheriff. Two months later deputies arrested Newman again, this time for taking a bribe. That charge stuck, but it took until the end of the year for Newman to exhaust his appeals and surrender his office. Despite their best efforts, the mine owners and managers were stuck with Newman for most of the strike.[9]

The miners finally struck in mid-June. Thomas Cornish, who had worked for Smith in Leadville since 1894, said he discovered that a non-union man by the name of Gregory had been working on the Garbutt Mine for about three months. On the night of June 19, 150 men from the Little Jonny surrounded the house where Gregory lived and demanded that he join the union or be put off the hill. Gregory refused, and a labor strike started the same day. A total of 968 miners struck, 375 of them employed by Smith-Moffat. While the strike was important to Smith, of far more interest to him was the fast-approaching 1896 presidential election. Once again, Smith was about to take a short leave from his allegiance to the Republican Party.[10]

Silver was far more important to Eben Smith than political party, and if restoring silver meant leaving the Republican fold, he was prepared to do it. Like Smith, a number of people were still upset over the repeal of the Sherman Silver Purchase Act as campaigning for the 1896 presidential nomination got under way. One Democratic candidate, William Jennings Bryan, a former congressman from Nebraska, heard their cries and spoke throughout the South and West on the issue. Bryan arrived at the Democrats' convention with broad support due to these speeches, and when he pledged at the convention that the country's pro-gold interests would not "crucify mankind upon a cross of gold," he erased all doubt as to who would get the party's nomination. The next day Bryan won on the fifth ballot, splitting the Democratic Party as the pro-gold Democrats walked out and nominated their own candidate, Senator John Palmer of Illinois. The Populists met two weeks later in St. Louis and also nominated Bryan but named their own vice presidential candidate, former congressman Thomas Watson of Georgia. The Republicans nominated William McKinley, former congressman and governor of Ohio.[11]

As soon as Bryan had the nomination, Smith began campaigning for him, writing to H. L. Horton that Bryan was a "new Moses." He took the "liberty to go into the prophesying business," predicting that if Bryan won the election and silver recovered "its rightful functions," the products of labor would increase in value, the thousands of idle men in the country would be put back to work, securities would see their value rise, and the country in general would become prosperous. Smith also saw a Bryan victory as good for Wall Street, which he thought would resume its old activity and become a place where men of energy and means could make a living instead of resembling "a neglected graveyard where the fences have tumbled down and cattle are roaming at will dirtying on what they don't otherwise destroy."[12]

Smith's love for silver was so deep that he even put a Bryan victory above selling stock in the Smith-Moffat businesses in London. Horton informed Smith that those in London interested in mining stocks hoped for a McKinley victory since they preferred gold to silver, but Smith was unmoved. He had been mining silver for more than thirty years, and one of his strongest desires was to see it make a comeback. In September, he wrote to Charles

Thomas, president of the 1893 Silver Convention in Denver, that a Bryan victory would ensure the employment of double the number of men already employed in Leadville. As Smith continued to proclaim his support for Bryan, the striking miners in Leadville became increasingly restless.[13]

With nearly all his miners out on strike, Smith shut down the pumps on the Bon Air and Penrose Mines on August 30, 1896. This move so disturbed local businessmen that a committee of them met with Smith and offered to pay to keep the pumps going, but he refused. Then, on September 21, both the Coronado and Emmet Mines (neither of them a Smith-Moffat property) were bombed. The bombings convinced Sheriff Newman to admit that he had lost control of the situation, and he asked the governor to dispatch National Guard troops to Leadville. The troops, under the command of E. J. Brooks—the same general in charge of them during the earlier Cripple Creek strike—arrived the same day. The next day, Thomas Shelton reported to Smith that striking miners in Leadville had set fire to some buildings, killed about thirty men, and wounded fifty more. Shelton wrote to Eben, "Leadville will be no good untill [*sic*] we have martial law." By October 2 the soldiers had restored some order to the area and stood guard at a number of mines in Leadville, including the Maid of Erin. The owners and managers, in an effort to resume production at some of the mines, brought in scabs from Missouri to work the properties. They continued to use them through early 1897.[14]

November 1896 proved to be a difficult month for Eben Smith. William McKinley was victorious in the presidential election, dashing his hopes for a silver revival, and he also lost another child. His daughter Nellie had married Charles S. Jeffrey and was living in Sacramento. Smith received a telegram from B. F. Jeffrey and Simeon Jordan on November 7, 1896, informing him that "Nellie is very low" and "we fear the worst." Eben and Emily left for California two days later, arriving on November 14. Unfortunately, they arrived too late, as Nellie died on November 10 at age thirty-three due to complications from an ectopic pregnancy. Her funeral was held the next day at the home of her half-brother Simeon Jordan at 1310 H Street, after which her body was sent to Oakland to be buried next to her mother and brother Samuel in Mountain View Cemetery. On November 18, Eben sent a telegram to a business associate, former US senator Stephen W. Dorsey, explaining that his absence from Colorado and his business affairs was due to the "illness and death" of his daughter, the third of his children to die.[15]

The situation in Leadville was just as tense when the Smiths returned as it had been when they left. The Little Jonny, which as a gold-producing mine continued to work through the strike, faced a number of lawsuits filed by people who, as Smith told his old friend Dillwin Parrish, were "blackmailing something" out of mine owners on a trumped-up charge. Smith was not as dismissive of the lawsuits to Thomas W. Goad, who was in charge of selling Little Jonny stock in England. Although confident that he and his fellow owners would win the suits, he was afraid the English would not understand the situation and, as a result, the price of the stock would fall. He also feared the lawsuits would hurt the Resurrection stock he and Moffat were attempting to sell to raise money to further develop the mine. Smith wrote to Goad, who was also handling the Resurrection stock in London, that he thought "very highly of this property and was it not for the great length of time it will take to open it up I should be opposed to selling any part of it." The enormous amount of money required to turn the mine into a working property prevented the stock from selling, but time would prove Smith correct as to the mine's potential. Never one to let a setback slow him down, in December Smith, Moffat, Sylvester Smith, and Lafayette Campbell hired E. H. Crawford to develop the recently leased Home Again claim in downtown Leadville. As with all the other mining properties he was interested in, Smith had great hopes for the Home Again. Strike or not, Smith was a mining man, and that was what he intended to do.[16]

Although the strike did not end until March 1897, by December 1896 the situation in Leadville was stable enough that Smith felt comfortable leaving on an extended tour through the South with Emily. While traveling, Smith counted on Robert Reid to keep him informed about his business interests and took time to keep Reid informed about the progress of the trip. The first stop for Eben and Emily was Galveston, Texas, where they arrived on December 23. Smith wrote to Reid that the people in Galveston were mostly old-time southerners who raised a lot of cattle. According to Smith, they were "small and of an inferior graid [*sic*] about on a par with everything else they undertake." He saw nothing wrong with Galveston, though, that "an injection of some yankee [*sic*] enterprise" would not fix.[17]

The day after Christmas, the Smiths arrived in New Orleans, and what Smith saw there prompted him to write to Reid that he had "told Mrs. S. they did not have the yellow fever near often enough and not half severe enough." According to Smith, the streets in the city were filthier than those anywhere else in America, and the common people were the most "dirtyest [*sic*], Godforsaken looking beings I ever saw huddled together." Smith was even disappointed by what he had heard were supposed to be the most beautiful ladies in the South, as the only ones he saw had "no arses," and in his opinion, "a double barreled shot gun would make any of them a pair of drawers." He went on to say that if Ben Butler, the general in charge of New Orleans after its capture by the Northern army during the Civil War, had lived a thousand years, he might have been able to regenerate the people there. Smith told Reid that he damned the South and all the people in it.[18]

Eben Smith's descriptions of New Orleans were very entertaining to the people in his office in Denver. Reid wrote to Smith that he was sorry the women's build was not as broad as Smith would have liked but that due to the warm weather in the South, he thought there would be no "great demand for double-barrelled guns to use for the purpose you mention." Norris Cone, in closing his letter that gave a detailed account of how all the Smith-Moffat Cripple Creek mines were operating, told Smith he must have had the wrong glasses on when looking at the women. Cone wrote that he thought the glasses Smith wore must have been "far sighted and if you will get a pair of near sighted glasses the legs will look better."[19]

In late December, Eben and Emily arrived in St. Augustine, Florida, where they stayed into January 1897. Eben found Florida to be a fairly pleasant part of the South, though Emily did not like the heat. However, not being near his businesses was beginning to bother him. At sixty-five years old, Smith was uninterested in taking it easy, and he did not know how much longer he would be able to endure "loafing," writing to Reid that "the seat of my breaches is getting thin and my hip bones are sore." Still, there was one benefit to his loafing, he told Reid, as he felt much rested since he was "not being compelled to listen to a lot of people who were eternaly [*sic*] after something without giving anything in return." After a few days in Florida, Eben and Emily headed to Richmond, Virginia.[20]

Smith did not find Richmond as objectionable as New Orleans, but it was also not as pleasant as Florida. As part of their sightseeing, the Smiths visited the Civil War battlefield of Seven Pines. Smith wrote to Moffat of his amazement at how the battlefield was preserved, particularly when he ran into an old soldier there who told him he had bought the land and kept it and made his money by showing tourists around. Smith expressed his belief that the government should be responsible for maintaining the property. By then, the Smiths had had enough of the South, and Eben reported to Moffat that they would soon be in New York and that he hoped to "hear from you all as they say down here."[21]

Shortly after his arrival in New York, Smith had lunch with "some of the richest Jew Bankers in New York," who told him that mines were going to be the next great "speculation feature" in New York and Boston. In his history of the Gilded Age in the West, Gene M. Gressley writes that by the 1890s, easterners were in fact finding it more profitable to invest in mining stocks than to attempt mining themselves. However, those investors wanted minimal risk, and having a skilled manager appeared to lessen that risk. In Colorado, Smith was considered to be among the best managers. Sensing that what he heard from the bankers was good news for the Victor, Smith wrote to Moffat that he wanted William H. Brevoort, an employee at the Victor Exchange Office at 66 Broadway in New York, made president of the Victor Gold Mining Company. Smith thought it made good sense to have a man with connections in that city in charge of the mine to make it more attractive to eastern investors. Moffat agreed, but Brevoort was president in name only, and the mine remained under Smith's control.[22]

When the Smiths returned to Colorado in February, the strike in Leadville was winding down. The state legislature, tiring of the situation, issued a series of suggestions for ending the strike that neither side followed. On February 9, Smith, weary of the union's antics, instructed another business partner, John Canning, not to hire any union men to work on the Smith-Moffat leases in Leadville. Eleven days later, he wrote to Cone urging him to give a job to any interested "soldier boys," who were boycotted by the union because of their service during the Leadville strike. Finally, in March 1897, the WFM, running low on money for funding the strike, called it off. The mines reopened slowly, and Charles Carnahan informed a committee of the Colorado legislature that output at the Smith-Moffat mines could be as much as 2,000 tons less than it had been before the strike. By mid-April, Smith had still not restarted the pumps on the Smith-Moffat downtown properties in Leadville, and he was in no hurry to do so. The end of the strike could not have come at a better time, however, because the crown jewel of the Smith-Moffat mines was about to start causing trouble of its own.[23]

On March 8, 1897, Eben Smith wrote to William Brevoort in New York that he was shutting off the $20,000-per-month dividends from the Victor only six months after telling H. L. Horton that he thought the mine would be paying dividends "indefinitely." Smith explained that the company needed the money to pay for a new air compressor to run machinery at the mine as well as for other improvements. Smith carefully made clear to Brevoort, who was to inform the stockholders, that "the fact that we do this does not materially lessen the value of the property or its prospects." Nine days later, more trouble came as the officers of the Phil Sheridan Mining and Milling Company brought accusations that the Victor tunnel had crossed onto their property. Horace Tabor was one of the owners of the Phil Sheridan, and in mid-March, Smith wrote to Cone that he should allow Tabor to inspect the Victor tunnel. Tabor hoped the Phil Sheridan might help restore his lost fortune, but Smith-Moffat won the lawsuit and gained control of the property, renaming it the Venture Gold Mining Company.[24]

As Smith continued to direct improvements at the Victor through April, he also started selling stock in the Granite, a mine he bought without David Moffat's participation. In an effort to drive up the price of the stock, Smith told Moffat he wanted him to appear to be one of the owners of the mine. Moffat wrote that he understood why Smith wanted that and that if asked, he would not deny being one of the owners. However, the reputations of both Smith and Moffat were not enough to overcome the Granite's problematic reputation, and Smith's difficulty in selling stock was only the beginning of a great deal of trouble that would plague the mine.[25]

Selling stock was one of, if not the, principal ways Smith and Moffat made money during this period. At the beginning of April, business associate and former Arkansas senator Stephen W. Dorsey suggested that Smith should go to London to assist in selling stock in the Florence and Cripple Creek Railroad and the Gold Reduction Works of Colorado. Smith immediately set out to collect introduction letters from influential people, such as one from John Searles, which stated that Smith was in London for the "purpose of interesting English capital in some very valuable properties in Cripple Creek and Florence." In early May, he wrote to William Brevoort that he would "be mortified to death if we do not succeed in doing some business after reaching there." To enhance his chances of success, Smith asked Brevoort for suggestions on what to wear. He reported that he was having a new tuxedo coat and overcoat made, and he told Brevoort that he had a nice "mixed gray" suit that would be a good "boss" suit for both the trip and doing business in London. Brevoort approved of his employer's choice of clothing and could think of nothing to add.[26]

As Smith continued his preparations for the trip, he fired off the usual round of instructions to his people at the Smith-Moffat mines, telling them to keep busy and to report to him frequently. One of the final instructions Smith gave David Moffat before leaving was to buy the property onto which their Gold Standard tunnel in Cripple Creek had encroached. He reminded Moffat to be sure to have someone else buy the property for them, though, because there was "already such a hue and cry about our various operations." Smith used to fight these encroachment lawsuits, but as he and Moffat grew wealthier, that changed and it was simply easier and cheaper for them to buy any property onto which their mine shafts crossed. Smith also ordered the pumps on the downtown properties in Leadville restarted, much to the dismay of Moffat, who saw nothing in it for Smith-Moffat except a $12,500 loss.[27]

After a short stop at Liverpool to shake off their sea legs, the Smiths arrived in Paris on May 20. In Paris, Smith found a number of reports on the mines from Robert Reid awaiting him, much to his pleasure. What was in the reports, though, was not so pleasurable. In particular, he still thought the expenditures at the Victor were far too high, and he admitted to Reid that he regretted having started the pumps in Leadville. Not taking the time to indulge in recriminations over this move, Eben and Emily arrived in London on May 30, but not before letting Reid know that Eben expected him to write at least once a week with news of the mining properties.[28]

By the time they arrived in London, Emily said the trip had been her last sea voyage except for the return home. Eben wrote to Reid that he had heard women say that about having babies and then go on to have more, so he was taking her declaration lightly. The Smiths arrived in London just as Queen Victoria's Diamond Jubilee was getting under way, much to the dismay of Eben, who did not like crowds. Still, they were going to try to get tickets to the "show," as he called it, but if unsuccessful, they planned to go to Paris after a week or ten days in London. He had no complaints about their rooms in the city, which he said were fit for the queen herself. However, the trip was still business-related, which was where Smith focused his attention. Longtime business associate Dillwin Parrish introduced Smith to the leading men of the Gold and Silver Exploration Company in London as the first step toward interesting English investors in the Smith-Moffat properties. After filling Reid in on these developments, Smith asked him to start sending his daughter Cora $50 a month, as she was running short on pin money.[29]

The Smiths arrived in Paris once again in mid-June, where they spent a few days while Eben decided who to work with in London. Smith finally decided that Parrish was the right man, and he and Emily returned to London. Parrish and Smith met again with the men from the Gold and Silver Exploration Company on July 8. Smith went into the meeting with an optimistic attitude, but, as he informed Moffat in a letter that day, his mood changed drastically as soon as the meeting got under way. The men of the Gold and Silver Exploration Company insisted that an expert of their choosing examine the railroad and mill properties, at Smith and Moffat's expense, and that they be given full control of the property. Unwilling to give up control, Smith put on his hat and told them goodbye, ending the business part of his trip.[30]

The meeting was a setback to the Smith-Moffat plans, but Eben did not let it rob him of his time in Europe. In mid-July, he wrote to Reid that he was having a good time and "not sitting around like a bump on a log." Although he and Emily were not seeing eye to eye about what to do on the trip, he did join Emily and another lady on a trip to Scotland and Ireland. Eben did not record any details of this trip, and by mid-August the Smiths were back in Paris, a city Eben was very "fond" of and that he one day hoped to spend more time exploring. Eben reported to Reid that Emily had "gotten down off her high horse" and was acting "quite rational," so overall, he was now enjoying the trip. The Smiths left London at the beginning of September, and after arriving in New York on September 10, Smith asked Reid to "remember me to all the boys," as he had done many times during the trip. From New York, the Smiths traveled to Pennsylvania, where Eben spent a week visiting his family before returning to Denver in early October.[31]

Eben Smith's return to Colorado in October 1897 signaled the end of many of his Leadville mining interests. Smith and Moffat withdrew from the Leadville Pumping Association because Smith believed the downtown properties the two owned were nearly worked out. He also abandoned all their leases in the downtown area, including the Home Again, worked for less than a year by E. H. Crawford. Charles Limberg told Smith that he "exceedingly" regretted this decision, but Smith did leave all the machinery erected by Smith-Moffat on the properties and agreed to sell or rent it for the right price. Smith would come to regret this decision, but the condition of the mines at the time convinced him that it was the right thing to do. He still had faith in the Resurrection, the Maid of Erin, the Henriett, and the Little Jonny and continued to work those Leadville mines without interruption.[32]

Although Smith-Moffat's Leadville properties were showing some signs of aging, Smith did not let the small drop in income caused by that fact influence his personal life. He was still one of the richest and most powerful mining men in the state, and his and Emily's lifestyles reflected it. Emily's passion for buying jewelry was almost as strong as her husband's passion for buying mines, and she indulged it at the Bohm-Bristol Company in Denver, where she bought numerous rings, brooches, and necklaces. The Smiths also bought household items such as flatware and cut glass from Bohm-Bristol. Another of Emily's favorite stores was the Daniels and Fisher Dry Goods Company, where she and Eben bought clothes for themselves and toys for their grandchildren. Much to their pleasure, Frank and Josephine added two more grandchildren to the family in the late 1890s: Melvin Hill on May 13, 1898, and Frank Leonard on September 28, 1899.[33]

When Smith regained his fortune in the late 1880s, he decided that "he and his children were going to live comfortably for the rest of their lives." Eben and Emily were becoming increasingly interested in living comfortably, and in May 1898 Eben made sure they also could travel comfortably. That month he bought a new Baronial Brougham carriage, painted green and monogrammed "E.S." on both sides, from A. T. Demarest and Company in Denver for $1,485. To pull his new carriage, Smith had a choice of three fine horses—Mary, Ben, and Sci—in his carriage house. To further ensure his family's comfort, Eben decided that a man of his standing needed a country house. On July 1, 1898, he paid $45,000 for Estamere, a house in Palmer Lake, Colorado. The house included a billiard room, a smoking room, a reception room, and a solarium on the third floor. Smith planned a series of extensive improvements to the house, the biggest being the addition of a northeast wing for the use of the Smith children. Frank oversaw the work for his father.[34]

While Smith was more than happy to spend money on houses, carriages, and jewelry for his immediate family, he did not neglect his charitable responsibilities. In January 1898, Eben agreed to have a house built for his

FIGURE 5.1. Estamere in Palmer Lake. Louis O'Kane Collection.

sister Nancy in Union City, Pennsylvania. Although Eben initially wanted to spend only $2,000 on the house, he finally agreed to spend $3,000 because, as he told his cousin Melvin Smith, who was to oversee construction, "there is nothing like pleasing the women in matters of this kind." Smith also helped the wife of his former stepson Simeon Jordan, who was desperate to get her daughter out of Cripple Creek to be treated for curvature of the spine and severe abscesses in her head, by sending her $250. After receiving the money, she wrote to Smith that she could not "find words to express my gratitude to you for your kindness. But please except [*sic*] my heartfelt thanks." Even though the Jordans had not been a part of Eben's family for more than thirty years, he was still willing to help them.[35]

While charitable requests and offers of mining properties for sale dominated Smith's mail, there were those who wrote to him with other, perhaps less noble, purposes in mind. In June 1898, Smith and Moffat received a letter from J. S. McGinnis and Company. McGinnis wanted the men to help him "secure a place" for a saloon in the town of Goldfield, which had a "brilliant" outlook for such a business. Smith enjoyed a drink now and then, so he may have been inclined to help McGinnis, but he was also critical of those who drank to excess. It is unclear if he or Moffat assisted McGinnis in his quest, but a little more than a year later, their Gold Knob Mining Company donated two lots in Goldfield to help build the Methodist Episcopal Church. Regardless of whether the outlook for heavenly salvation was as brilliant as that of the saloon in the eyes of Goldfield's residents, Smith did his part to help.[36]

In December 1897, the Victor resumed paying dividends, once again making the stock attractive to investors. As a token of gratitude for his able management of the mines, David Moffat had an elaborate gold watch made for Smith using ore from one of their mines. Smith's large personal fortune enabled him and his family to live a luxurious lifestyle, and he continually improved it over time. However, there were signs at the beginning of 1898 that Smith-Moffat's luck was not going to hold forever.[37]

6

"Not a Pound of Ore in Sight"

In February 1898 William H. Brevoort, president of the Victor Gold Mining Company, traveled to France in an unsuccessful attempt to sell more Victor stock. Rumors of war between the United States and Spain circulated throughout France, and on February 14, 1898, Brevoort wrote to Eben Smith that in France, everyone was saying "damn the bulldozing arrogant Americans," a sentiment that did not bode well for the sale of American mining stocks. The French were particularly upset with the threat of war, in Brevoort's opinion, as the "frog eaters" owned "lots of Spanish bonds," the value of which would sink if a war broke out between the United States and Spain. Smith had also asked Brevoort to see what he could do to sell Granite stock in France, but he was unable to place any of it by the time the United States declared war on Spain on April 25, 1898.[1]

The war with Spain, which lasted from April to August 1898, completely shut off stock sales in Europe. With this market unavailable, Smith turned his attention to other stock market matters. He helped incorporate the Denver Stock Exchange and also served on its board of directors. He was also paying

DOI: 10.5876/9781646421794.c006

more attention to buying stock. In 1898, he invested in the Golden Circle Railroad Company, located in Cripple Creek, buying $22,800 in stock and $22,800 in bonds. As with all the other business ventures Smith invested in, he hoped the Golden Circle would prove to be a moneymaker for him.[2]

Smith also felt freer to inspect his mining properties more thoroughly during this time. In July, he traveled to Leadville, where his first stop was the Resurrection. There he found large quantities of low-grade ore and made a note to himself that he thought it "advisable to consentrate [*sic*] ore before marketing." Overall, though, he was happy with the way his son-in-law Charles Carnahan, who still lived in Leadville part-time, was managing the mine. The next day he visited the Bi-Metallic Smelter and "found all pretty snug and prosperous." Smith left Leadville satisfied by what he had found.[3]

Smith was also happy with Norris Cone's management of the Cripple Creek mines and left him alone to run them as long as he received regular reports on the properties. There had been no further trouble with the Victor Company's ability to pay dividends, but there was still trouble at the mine. Cone wrote to Smith on September 19 that workers had struck water in the Victor shaft, but he had plugged it up as best he could and was working on the stations on levels eleven and twelve of the mine. Smith occasionally sent reminders to Cone to keep expenses down, among other things, but for the most part he was happy to let Cone do his job without much interference.[4]

While everything appeared to be business as usual, there were signs of problems looming. In mid-1898, Smith found it necessary to loan money to two of his mining companies to help them meet expenses. Smith had loaned a great deal of money to the Gold Standard Mining and Tunneling Company in 1894 and 1895, but the rest of his mines had been able to pay their own ways. This changed, though, in July 1898 when conditions forced him to loan $1,000 to the Granite Gold Mining Company and $3,000 to the Gold Knob Mining Company. This money kept the mines in operation, but it was clear that their productivity was declining.[5]

With extra time on his hands, Smith also began hunting for new mining properties, dropping his remaining objections to mining outside Colorado as he searched. Smith soon focused his attention on Oregon, and in December he traveled there to see the mines in the state for himself. He passed through Utah on his way to Oregon and was so impressed by what he saw there that he hired A. J. Underwood to look into a number of properties. Within a year, Smith-Moffat became owners of the Carisa and Spy Mines in Utah, while in Oregon he leased the Granville and Hunkadora Mines and bought a mine near Placer, Oregon, that would eventually become one of the biggest Smith-Moffat properties outside of Colorado. The mine so awed Smith that he quickly renamed it the Victor Junior, over the opposition of Frank Sutherland, its new superintendent. When Smith returned to Colorado in early 1899, he was content to let Underwood and Sutherland manage the new mines as long as they sent him regular reports.[6]

While the Colorado mines were showing signs of trouble, the Denver Mine and Smelter Supply Company prospered. When a branch office opened in Mexico City in 1898, Smith became president of the Mexico Mine and Smelter Supply Company as well. His son Frank, accompanied by his wife, Josephine, and son Eben, visited Mexico City in May 1899 to inspect the operation. William "Guillermo" Bushnell, who worked in the Mexico division, was especially pleased to see them and proudly reported their visit to Eben in one of his regular reports. This was some of the last truly good news of the year.[7]

June 1899 turned out to be one of the most unpleasant months of Smith's life, both personally and professionally. Early in the month, Smith received a copy of a letter his wife sent from a hotel in Glenwood Springs, which suggested that he had admitted to being intimate with Grace Ripley of Colorado Springs after her marriage to miner Raymond Ripley. Emily, who suspected her husband of having had other affairs during their marriage, had apparently written a few of these letters before going to Glenwood Springs in May, but Eben wrote to lawyer John Deweese that he thought he had ended the onslaught by telling Emily that she was putting herself in an embarrassing position. Eben was "astounded" to learn of this latest letter from Emily, and he emphatically denied the charge, telling Deweese that he had had nothing to do with Mrs. Ripley since her marriage.[8]

According to Eben, Ripley planned to divorce his wife and was prepared to name him as the co-respondent in the case, with Emily's statements as evidence. Eben told Deweese that if forced to take the stand, Ripley's attempt would end in failure. Interestingly, given the personal nature of the letter to Deweese, although it is handwritten, it is not Eben's handwriting. Smith finally disposed of the Ripley affair in August 1899 when he made a contract of settlement with Mrs. Ripley for $2,250. Another lawyer, Samuel Berry, later tried and failed to involve Smith in a lawsuit against Mrs. Ripley's lawyer, as she claimed he had cheated her out of some of the money.[9]

In April 1899, Norris Cone married for the first time and, after a honeymoon in New York, settled in Denver. In early June, he fell ill with meningitis and died unexpectedly on June 15, 1899. Cone's death was a huge blow to the Smith-Moffat operations in Cripple Creek, as under his management the Victor alone had produced more than $2 million in ore and paid out $1.155 million in dividends. By the time of his death, however, Cone had run into some serious problems at the Victor. The previous November, after reporting on the installation of new drills on levels eleven and twelve of the mine, Cone told Smith he was having trouble locating the new vein miners had been following. At the time, he thought it had either gone perpendicular or crossed with the old vein. Six months later, Cone still could not locate the vein. After Cone's death, Eben Smith personally took over all the Cripple Creek mines, but he was no more successful than Cone had been.[10]

While Smith and his family spent their first summer at Estamere in Palmer Lake, he took the train to Denver each morning to attend to business. After Cone's death, Smith ran the Cripple Creek mines "with a view to bettering the conditions." This development especially pleased Moffat—who had done some "pleasant things" for Cone after his marriage—as he thought Cone had not "properly" attended to Smith-Moffat affairs for some time. Regardless of Moffat's happiness about the change in management, Smith had his work cut out for him. The Gold Knob, Arvilla, and Victor Mines were all losing money; and Smith closed them until he could figure out how to restore them to prosperity. Smith also closed the Golden Cycle until the smelters reopened. At the Anaconda, Smith dismissed what he described as the old men working in the mine and put a new man from Leadville, who he described as a "cracker jack," in charge. Smith wrote to Moffat that he thought the Anaconda was "in a fair way to come out all right" and that miners were taking ore out of the Roxana that looked "quite promising," which was perhaps the best news of all as far as the Cripple Creek mines were concerned.[11]

The situation in Leadville was not much better; miners were taking ore out of the Resurrection at the rate of 100 tons per day. In September, Smith leased all of the Maid of Erin property, where production had slowed dramatically, to his son Frank and his associates. Then, in late October, Smith learned that he had been badly mistaken when he withdrew from the

Leadville Pumping Association and let Smith-Moffat's leases on the Home Again and other downtown properties lapse in 1897. Workers had opened a new ore body, thought to be "one of the largest and most valuable ever . . . in Colorado," on the property. Smith, David Moffat, and John Campion viewed those who invested in the Home Again Company after they withdrew as people investing in a "forlorn hope," but with the discovery of the new ore, the *Rocky Mountain News* commented on the "lack of foresight" Smith and his associates had shown. This was one of the few times a major newspaper criticized Smith, even though it was lighthearted criticism.[12]

By the end of October, the lost vein at the Victor remained missing, and Smith feared the vein had shifted onto the property of the neighboring Isabella Mine. Two weeks later, on November 14, five men working on the fourteenth level of the mine got a dose of powder smoke. By the time mine employee F. T. Osgood wrote to Robert Reid informing him of the accident later that day, rumors in town purported that two men, Thomas Jones and Robert Conners, had died that evening. Osgood expected a third man to die, while two others were reportedly very ill. Osgood thought it was amazing that all the men were well enough to go home immediately after the incident instead of being "overcome at once as is always the case." That same day, Eben Smith wrote to Lee Wood, now in charge of the Gold Knob at Goldfield, giving more details of what happened. Friends saw Thomas Jones as late as 7:00 p.m. at a dance with his wife, but by 11:00 he was dead. William Dodsworth, the man Osgood thought would die, was likely to survive. The two other men, French and Fairbanks, could not be located, so their condition was unknown, although rumors suggested that they were dead. It was with great relief that Smith reported to Wood that "the men that are sick do not hold the Company responsible in any manner." Smith's sense of relief grew even more when he learned that French and Fairbanks had recovered.[13]

Six days after the powder smoke incident at the Victor, Smith put aside his troubles and left for Mexico City to examine a number of potential mining properties. On November 1, 1899, Smith hired Edward Thomas to find veins containing gold, silver, and copper in Sonora, Mexico, and Arizona Territory. Smith expected to be gone for about three weeks, and he wrote to Lee Wood at Goldfield to "keep things humming while I am absent, and be ready to render a good account of yourself when I return." Had the Smith-Moffat mines been in better condition, Smith would have instructed Wood to write to him in Mexico, but with the situation as it was, he waived that requirement.[14]

As Smith searched for new mines, he sounded like those who had written to him over the years offering the most impressive mine ever. On November 27, he wrote to Thomas Goad from Mexico that he expected to get an option on a mine in British Columbia, the shaft of which was down to a depth of 210 feet. At 70 feet, the vein was reportedly 7 feet wide and the ore was running $35 per ton; at 125 feet, the vein was 12 feet wide with ore running $50 per ton; at 210 feet, the vein was 21 feet wide with ore running $70 to $120 per ton. Smith wrote to Goad that these figures sounded like "a great big fairy tale, and may turn out so, but as our source of information is very reliable and of the best, it only remains to be seen whether the people themselves are being deceived." He expected to have the option within a few days and would then have the mine inspected. It must have turned out to be a big fairy tale because Smith never again mentioned the British Columbia mine.[15]

Displeased with what he found in Mexico, Smith returned to Colorado and the bleak situation he faced at most of his mines. He ended the year on a high note, though, when Smith, his attorney Tyson S. Dines, Robert Reid and his brother W. D. Reid, and O. P. Poole incorporated the Denver-Galena Zinc Mining Company for the purpose of mining zinc in Colorado, Kansas, and Missouri. It was one of the last new companies he incorporated in Denver, as the rest of his time was spent winding up the affairs of the old ones.[16]

In January 1900, Smith and Moffat sold the Gold Knob Mining Company for $132,500, of which Smith received $34,015. Smith also announced that he was interested in selling the Resurrection Mine in Leadville. W. W. Allen wanted to buy an option on the mine, and he told Smith on January 20 that he would determine in the shortest possible time whether he could handle the property. Eben was interested in Allen's proposition, but negotiations stalled over the price, and several months later the mine was still in Smith-Moffat hands, much to their relief.[17]

In February, Smith again took up the matter of the Victor Mine, where the vein remained lost. On February 2, he wrote to Moffat that he wanted to sell the Victor to the Isabella Gold Mining Company, onto whose property he feared the Victor vein had drifted. Smith argued that this would give the French and eastern stockholders "a better show for their money," pay off the company's $125,000 debt, and pay him $2.50 to $3.00 a share. Smith was especially anxious to do this, he wrote Moffat, "in view of the fact that there is not" and might never again be "a pound of ore in sight in the mine." The plan fell apart, as Smith informed Robert Reid on February 23, because Williams of the Isabella refused to pay more than 50 cents per share. Moffat balked at selling for such a low price and, much to Smith's dismay, was ready and willing to spend more money to try to find the vein again. Moffat was in the process of putting together a scheme to solve the financial woes of the Victor Gold Mining Company.[18]

Frustrated with conditions in Colorado, Eben and Emily decided to travel to California in early February. Emily's sister, Sarah Rundel, who had moved in with the Smiths about this time, joined them. As usual, Smith fired off a last round of instructions before leaving. Of particular interest to him was the Granite Mine, where miners were having trouble with the hoist, and he instructed Robert Reid to give up working on it if new, better equipment could be obtained for less. Above all, he told Reid to "keep me fully posted about all the various matters in which I am interested." By February 11, the Smith party was in Oregon, where Eben stopped to inspect the Victor Junior. He was generally pleased with what he saw, but he wrote to Reid "it is pretty hard to determine at this time how the Victor Jr. will turn out in the long run." The fifth level of the mine, which was the deepest level at the time, did not look good to him. The south vein was very rich and well defined and looked better in the bottom of the level than in the back, but he was worried because it was also very small. Despite these problems, Eben still thought he would get his money out of the Victor Junior and that it might even turn into a good mine.[19]

The Smith party arrived in California on February 17, where Eben was annoyed to find no reports from Reid waiting for him. He wrote to his secretary, asking "has your hand lost its cunning that you are unable to write?" Reid immediately sent a detailed report, but there was not much in it to please Smith. As the Smiths traveled around California, the Granite continued to receive a great deal of Eben's attention. On February 27, he wrote a detailed description of how the new armature for the motor he wanted would work and then started in on the hoist. Frank Sutherland, then at the Granite, wanted to fix the old hoist, but Smith objected, writing to Reid "I don't propose to spend any more money on that machine except by the advice of competent electric engineers. It is too expensive to monkey with people that are guessing." Smith prevailed, and on March 27, workers began installing a new hoist at the Granite, although Smith was careful to instruct Reid to keep the men putting it in from working more than two shifts. Two major strikes had taught Smith a thing or two about labor relations, and he wanted to avoid any more trouble.[20]

As usual, business was never far from Smith's mind, even while on vacation. In March, he received an offer of $250,000 for the Victor Junior, which he wanted to accept. Smith forwarded this proposal to Moffat. After receiving no reply, he wrote to Reid that he figured that Moffat, as usual, would "think somebody is making something out of it on the inside and will probably refuse to sell it." Striking out on his own, at the end of the month Eben and his son Frank went into the oil business with G. A. Whiteford. Fletcher Jordan once described Jerome Chaffee as a man always awake to money-making possibilities, and by this time in his life, the same was true of Eben Smith. Searching for oil was a new departure for Eben. Although Whiteford spent years searching for oil in California and Oregon, he never found enough to make the venture profitable. However, Whiteford and Eben did maintain a good business relationship for the rest of Eben's life.[21]

At the end of March, the Smiths and Sarah Rundel arrived in San Francisco. They were not staying in the Palace Hotel as expected because, as Eben wrote to Robert Reid, they "couldn't find closets large enough for the ladies." California so charmed Eben and Emily during their visit that they found themselves thinking about moving there, but the thought of being so far from their children and grandchildren seemed to make doing so impossible. One thing that Eben found particularly attractive about California was the fact that the Irish had not taken control of politics there. He wrote to Reid that, unlike in Denver, the "American man" managed to run the political scene and would "till some damned Tom Paterson [*sic*] SOB starts a paper and tells them how they are being abused and deprived of the privileges guaranteed by the Constitution of the U.S.[,] something they never heard of."[22]

The progressive movement, which focused in part on business regulation, had been gaining strength in Colorado since the 1890s. As Smith saw his mining empire beginning to decline, he may have blamed part of the decline on this movement. He wrote to Reid from California that there was too much of a "desire to stomp on anyone who had a dollar in Colorado." He thought that "if a lot of us who have built up that country would fold our tents and leave it[,] the Yellow Dog element would turn coattails and drown each other which to me would be a happy result." Thomas M. Patterson, a Democrat, served one term in the US Congress in the late 1870s before returning to Colorado and eventually buying the *Rocky Mountain News*. He used the paper to "crusade" for progressive reforms and against large corporations, in addition to promoting the interests of working people, all of which were of little interest to a mine manager like Eben Smith. In the 1900 election, much to Smith's dismay, the state legislature made Patterson Colorado's newest senator.[23]

While Smith was frustrated both politically and in his mining business, his son Frank was trying make his own way in the mining world. In early March, Eben wrote to Reid, authorizing him to loan Frank $22,000 to buy the silver-producing Pride of the West Mine in Silverton, the title of which was involved in litigation. Five days later he wrote to Arthur Wilfley, Frank's partner, that he would put up the entire $110,000 first payment on the Pride of the West if Wilfley extended the option Smith had on the property until settlement of the title issue. Smith promised that he would take action at the earliest possible time to obtain a settlement, as he had "some pretty strong pulls in Washington," if Wilfley took the deal. This obviously worked, as he and Frank were soon the owners of the mine.[24]

The Smiths and Sarah Rundel returned to Colorado in late May or early June 1900, and for the first time in a long while, there was some good news. On June 13, Smith wrote to David Moffat that workers had discovered a new vein at the Resurrection, and because of this discovery, Charles Carnahan did not want to sell the property. Smith quickly canceled the proposed sale to W. W. Allen. The next day, Smith received a letter from J. M. Swain, whom he and Carnahan had hired to search for mines in Mexico. Swain found what he thought was a good copper mine and wanted Smith to get it before the "Gugenheimers" [*sic*] (the Guggenheim family) did. Swain reported that the Mexicans who were working the surface of the mine made enough money to build houses and pay their expenses, keeping them in "peon luxury." Despite Swain's praise of it, there is no evidence that Smith bought the mine.[25]

While the Smiths toured California, Moffat was in Denver putting his plan for the Victor into action. On March 22 Smith, as vice president of the Victor Gold Mining Company, supposedly signed several promissory notes totaling $151,000 from the Victor to the First National Bank of Denver. Moffat, who was president of the Victor, issued the notes as president of First National

Bank. It is possible that Smith signed the notes and backdated them later, but the subsequent chain of events relied on the March 22 date. Moffat's plan, put into action later in the year, was to sue the Victor, as president of the bank, for defaulting on the notes. This, in turn, required the county sheriff to seize the property and auction it to pay the company's debts. Moffat planned to buy it back, enabling him and Smith to start over with a debt-free property.[26]

News of the Victor's predicament spread quickly. At the end of March, V. Hurtado, a spokesman for the French stockholders, wrote to William Brevoort that he understood the Victor was in bad shape financially, a fact that was "painfully impressed" by an earlier letter from Brevoort. Hurtado hoped the "Denver Bank (i.e. Mr. Moffat, who had made a big profit by the sale of his stock), may be asked to be lenient, and not demand a pound of flesh." More than anything, Hurtado wanted a reorganization plan for the company that protected the original stockholders, which was what Smith was trying to do with his plan to sell the Victor to the Isabella Company.[27]

After a quiet July, Moffat put his plan for the Victor into action in August. On August 15 he, as president of First National Bank, filed suit against the Victor for defaulting on the promissory notes Smith had signed in March. According to the complaint, the company paid the interest on the notes through July 5, at which time payments stopped. Moffat wanted $151,000 plus the interest that was due on the notes since July 5. On September 6, a judge awarded Moffat and the bank $155,000 and ordered that the Victor be sold. The day after the judgment, Smith wrote to William Brevoort that he and Moffat wanted to organize a new company and raise $300,000. Of this money, $155,000 was to pay off the debt of the old company while the rest went into a new working fund. As time passed, Smith began to fear that someone might outbid Smith-Moffat at the October 10 auction and wrote to Moffat that although he wanted the mine back, he did not want to go too high with their bid. With the auction not taking place for a few more weeks, the only thing Smith could do was wait, worry, and run his other mines.[28]

After an agonizing wait, the October 10 auction arrived, but Clayton Dorsey, acting as attorney for Smith and Moffat, had the sale delayed until November 1 and 2. Inevitably, Smith-Moffat had to inform the stockholders about what had happened, which they did in mid-October. Brevoort and Reid sent a letter to all stockholders about the judgment and forthcoming auction.

Hoping to ease some of the stockholders' discomfort, they were careful to point out that from February 1893 to December 1898, ore sales at the Victor amounted to $2,161,186, of which $1,155,000 went to dividends—suggesting they had made a good return on their investments. On November 1, Brevoort wrote to Reid that those stockholders with whom he discussed the Victor's situation "did not hold Mr. Smith or myself morally or legally responsible, notwithstanding the investment has been most disappointing." At the auction on November 2, Dorsey successfully bid on the property for Moffat, and Smith and Moffat were once again the owners of the Victor Mine.[29]

With the Victor out of commission, Smith relied on his other mines to hold his attention. In looking at the Carisa Mine in Utah, he found that the value of the ore taken out of the mine was dropping. Smith wrote to A. J. Underwood in September to find out why that was the case, suggesting that it might be advisable to go to another smelter. Underwood's answer explained in detail why the value had dropped and went on to explain why he chose J. C. Griffiths to sample the ore instead of American Smelter. The decision was a logical one that would have pleased Smith—Griffiths charged $14.10 per ton while the American Smelter charged $17.00. As one who liked to keep smelting costs down, Smith approved of Underwood's decision. Still, he was not completely satisfied, and Smith continued to press Underwood about the ore. At the beginning of October, Underwood wrote that "a very <u>conservative</u> estimate of our ore bodies would be $100,000.00 net." Smith, finally satisfied, left Underwood to his business.[30]

The Granite Mine continued to cause trouble for Smith as he tried to put the Victor Mine back together in early November. Smith informed Superintendent Lee Wood that a man named Don Parks was making trouble, telling people who were friendly to Smith that grading being done at the Granite for the new engine was actually on the grounds of the neighboring Terresa Mine, on which Parks had a lease. Parks did not want anyone to say anything about the encroachment until the engine was actually in place. Smith asked Wood to look into the matter, saying "this man Parks is as dirty a man as ever went unhung, and why he should take such a position is something I can't understand." Smith surely did understand why Parks would say such a thing, since he would have been in line to receive money for damages if the engine was on his property. In the end, whatever action Wood took had the desired result, as Parks caused no more trouble.[31]

Lurking behind all of Smith's business activities during this time was a decision he had made while traveling through California. He had decided that it was time to begin a slow withdrawal from the mining business, and he made his first move toward this plan in the summer of 1900 while at Estamere, when he instructed "Guillermo" Bushnell at the Mexico Mine and Smelter Supply Company to sell all his stock in that venture. His next move came in early September, when he wrote to Richardson, Hill and Company in California, introducing Frank Sutherland. Sutherland had devised a dredging plan for a property near Oroville, California, in which he was trying to interest investors. Smith wrote to Richardson, Hill that he and Moffat were going out of the mining business as quickly as possible. At sixty-eight, Smith felt he was "getting too old, and when I am not able to take personal charge and cognizance of such business I do not care to continue it." However, he had no problem recommending Sutherland's plan, and if he were not feeling his age, Smith would have been eager to have been a part of it.[32]

Despite his pledge of retirement, Smith found himself having difficulties completely extracting himself from the mining business. On November 5, he and Moffat bought the Little Don Mining and Milling Company, located in the Cripple Creek district, for $25,000—each of them purchasing half of the company's 625,000 shares. Earlier in the year, Smith had also become president of the Independence Consolidated Gold Mining Company in the Cripple Creek district. The company was in a shaky financial position, and in August, Smith had to loan it $1,000. In December Reid, who was also involved with the company, asked Smith for additional funds. Smith told him to shut down the mine if it could not support itself and that he did not want to give it any more money, as he felt "I have had my legg [*sic*] pulled more than sufficiently on that account."[33]

The year 1901 started very quietly for the Smiths. Eben reported to his office in the Equitable Building on most days, but little at the mines required his constant attention. Having sold or shut down most of his Colorado mines, he had little left to do as far as they were concerned. The mines in Utah and Oregon were in safe hands and received less than the usual amount of attention as well. Smith still expected to receive the usual reports, and he kept close track of the assay reports. The slowdown did not hurt his wealth, and he continued to be very charitable. In 1900 alone, he had loaned the University of Colorado $5,000 when the school ran into serious financial trouble, his daughter Cora's friend C. M. Ritchie $500 to save three lots she was about to lose through "defraud of two villains," and John Woodbury, the caretaker of Estamere, $500 to pay off some lots he owned in Palmer Lake and to build a cottage or a store with living quarters above it.[34]

Eben Smith's less than rapt attention did not mean there was no good news from the mines, although it was rare. In January 1901, newspaper reports described the Pride of the West Mine as a "new bonanza." Owners Eben and Frank Smith and Eben's brother-in-law C. W. Denison had leased the mine the previous year to H. Kennedy, who uncovered the main vein and a "rich pay streak." Unhappily for Kennedy, he let the lease lapse, returning the mine to the control of the Smiths and Denison. At the end of January, Eben and his associates were in the process of extending the lower crosscut to the vein and building a large concentrating plant to handle the large amounts of low-grade ore from the mine. The *Denver Times* predicted that the Pride of the West would become "an important producer during the next few months."[35]

In early February, the Fortune Mining Company in Leadville filed a lawsuit against the Resurrection for the routine claim that the Resurrection shaft had crossed onto Fortune property. Officers of the Fortune sought $15,000 in damages, a relatively small amount, but Charles Carnahan took no chances with it. He gave himself a mortgage on the mine to secure its indebtedness to Smith, Moffat, and himself; and he wrote to Smith that before they would let "those damned robbers have a cent we will foreclose and let them pay us over $300,000 and take the property or get nothing." Carnahan's plan worked, as the property remained in Smith-Moffat hands. At about the same time, Eben's son Lemuel, deep into a long struggle with alcoholism, went to work at the Resurrection.[36]

Two months later, Robert Reid reported to his boss that there had been a "very bad accident" at the Granite Mine, the worst in a long line of troubles at that property. The cribbing on the Granite dump had given way, and tons of ore slid down onto the property of the neighboring Monument Mine. The slide killed six people by the name of Smith (no relation to Eben) when it slammed into a small cabin they occupied below the dump, covered the Monument ground with Granite ore, and destroyed the water tank on that property. Immediately after hearing of the accident, Reid sent his brother, lawyer W. D. Reid, "to take all the steps necessary to protect us against damage suits." Reid reported to Smith that he thought the company could settle for a maximum of $10,000.[37]

The heirs of the six people killed still had not approached the company a week later. Reid was expecting to hear from them at any time, and "in a round about way" he had them approached and advised not to see an attorney but rather to "see the officers directly, and save being skinned by an attorney" if they thought they were owed damages. In reality, Reid and, by extension, Smith were probably fearful that an attorney would skin them and successfully sue for more than the $10,000 for which Reid thought they could settle. In addition to the family of the people killed, the company also had to deal with the Monument Mine, as ore from the Granite dump covered its property. Smith eventually reached a settlement with the Monument officers that was less than pleasing to both sides, but it ended the matter. Reid and Smith never mentioned any settlements with the families of the six people killed.[38]

In July, the Smith-Moffat interests received some good news from the Resurrection in Leadville. That month the Number One shaft reached a huge body of sulphides "of fine strength and value" below the 1,000-foot point. Workers extended the triple-compartment Number Two shaft below the huge ore deposits, giving the company a "most economical arena" for extraction, development, and drainage. These recent developments persuaded the *Denver Times* to announce that Smith's judgment of the mine "was faultless and by the later developments a great bonanza has in reality been resurrected." Smith, Moffat, and Carnahan were very pleased that they had been unable to reach a deal to sell the mine the year before.[39]

Still, by 1901 it was clear that the majority of Smith's business interests in Colorado were winding down. The Cripple Creek years rivaled his years in Central City for being the most active as far as mining was concerned. In Cripple Creek, he became involved in one new mining venture after another with almost dizzying speed, and according to the *Denver Times* his success there "proved to then skeptical Coloradoans and to all-wise, scoffing geologists of the period that the camp was a permanent one and worthy of the heavy investment of outside capital." Smith's many years of experience served him well in his ability to know whether a certain mining property would be productive, and he became very rich because of it. While he built a good part of his fortune through stock sales during this period and it did not hurt that he and Moffat were able to drive up the price of the stock and sell at the right moment, Eben Smith still found great satisfaction in running the mines, and he did what he could to ensure their continued productivity.[40]

It was obvious by 1901, though, that the industry Smith loved was forever changing. After 1893, silver miners in Colorado had to adjust to no longer having government support for their product, and the price never recovered to pre-1893 levels. In 1899, smelting operations throughout the country, including the Bi-Metallic in Leadville, had joined together to form the American Smelting and Refining Company. With that, the infamous trusts that defined business operations of the era had arrived in mining. Consolidation followed among the railroads, which, in turn, made it extremely difficult for David Moffat to build his road over the Continental Divide. The Western Federation of Miners became an increasingly dominant force in Colorado's gold and silver mines during this period, and by 1902, one-fourth of the union's local units were in Colorado. Smith's response was to withdraw.[41]

When Democrats and silver supporters combined to elect Charles S. Thomas—chairman of the Colorado State Silver League Convention in 1893—governor in 1899, Eben Smith made no comment to any of his associates. Smith's lack of attentive management during this period suggests that he was happy to let Carnahan and Reid handle more of the details that he himself used to look after. In addition, a decision he and Emily made shortly after the new year began kept him preoccupied. Realizing that their children and grandchildren were only a train ride away, Eben and Emily decided to move to Los Angeles. Eben was in good health at age sixty-eight, but he was convinced that moving to Los Angeles would prolong his life well into his eighties. Emily was insistent about the move, and Eben reportedly agreed to it after her "urgent solicitation." The search was on for a new house as the Smiths prepared to leave Colorado.[42]

7

"I Cannot Pass a Good Mining Scheme By"

In April 1901, Eben Smith made one of the biggest financial moves of his life, buying $1,124,000 in US government bonds, which he then deposited at the First National Bank of Denver. Smith also interviewed a number of architects during the month as he toyed with the possibility of building a new house in Los Angeles instead of buying an existing one, but he finally settled on a slightly run-down house on 4 acres at 1130 West Seventh Avenue in Los Angeles. He purchased the house in May, but Mrs. J. W. Winston, the owner, had until July 4 to move out. On June 1, Smith wrote to M. Welsh, his agent in Los Angeles, that he wanted to enlarge the library, living room, and dining room by adding on to the house in addition to building a separate structure to house a "billiard room and other amusement arrangements."[1]

As soon as Mrs. Winston vacated the Smiths' new house on July 4, workers redecorated, installed new floors, and built a stone-faced wall around the entire property. On August 6, shortly after work on the wall began, Welsh reported to Smith that the house would be ready by October and shortly after that wrote "you have one of the handsomest places in Los Angeles."

DOI: 10.5876/9781646421794.c007

FIGURE 7.1. The Smiths' house in Los Angeles with Eben's private casino on the right, ca. 1902. Michel Tritt Collection.

He went on to say that people who could see the house from the train as it passed were amazed at the transformation. The private casino, the building that housed Smith's billiard table and other games, was also finished by then. To furnish the casino, Smith bought a Carom table and a pool table from the Brunswick-Balke-Collender Company. There was also a conservatory, a large pond, and a Japanese tea house on the grounds of Smith's new estate. The *Los Angeles Times* called it a "palatial mansion."[2]

While work continued on the new house during the summer, Eben and his family went to Palmer Lake for their usual stay at Estamere. For the first time, the public was able to read what life was like for the Smiths at Palmer Lake thanks to the *Denver Times*. A string of fine horses and ponies, a retinue of servants, and "all the luxuries which characterize his place in town" accompanied Smith and his family to the "big, rambling, comfortable"-looking house. The rooms were large, cool, airy, and full of sunlight; according to the *Times*, Smith furnished Estamere like a "handsome town house," with expensive wallpapers, rugs, and furniture. Soft couches with even softer cushions occupied nooks throughout the house, and there were a large number of "cosy

FIGURE 7.2. Eben Smith with grandchildren Harold and Doris Carnahan at Estamere, early 1900s. Louis O'Kane Collection.

[*sic*] corners" for people to hide in. The prettiest, according to the *Denver Times*, was the Turkish nook, which was "fitted up with real Oriental draperies." The house also had two dining rooms—one for the children and one for the adults.[3]

A dozen "blooded" horses as well as three ponies for the children lived in the stables, and they had surreys, broughams, a big drag, traps, and tiny carts for children. The house was a happy place, and the Smiths were more than willing to share it. Guests came and went throughout the summer, including poor women and children Eben and Emily invited to spend time at the house to feel the "joy of country life." They also invited sick children to the house, where they were "given a glimpse of fairyland" within the gates.[4]

The public also learned of Eben Smith's impending retirement that summer, again thanks to the *Times*. In a story on June 2, 1901, the newspaper announced that Smith had decided to "dispose of most of his interests in

Colorado" and spend the majority of his time in California, though he would still spend summers at Palmer Lake. He said he expected to visit Colorado from time to time on business, but he did not foresee actively participating in business and was willing to dispose of his business interests, with the *Times* reporting that he would not "sacrifice any of his property, but when it can be disposed of to advantage he will not hesitate to get rid of it." Smith's retirement from active work in mining in Colorado, according to the *Times*, took from the field "a man who has been most prominent in the development of the mineral resources of the state during the past forty years." The news came as a shock to many of his business associates.[5]

The news of his retirement was the cause of some amusement for Eben. On June 10, longtime friend Bela Buell, having seen the *Times* article, wrote to his old friend about it. Smith could not resist taking the time to correct a few inaccuracies in the article. The article reported that Smith was seventy-three at the time. He wrote to Buell "you are well aware that I am only 69." He also took issue with the fact that he was reported to be worth $10 million, telling Buell he believed he was worth only $1 million. Overall, though, he was fairly pleased with the article, as "this is about as close as the news papers of the present time, and particularly in Denver, get to the truth of matters."[6]

Smith wrote to longtime associate Morris Estes on August 1 that he was "trying to get out of business but do not seem to succeed very well, as I cannot pass a good mining scheme by." By then he had already hired Charles Mater to investigate the recently leased Crown Point Mine in the Crescent Mining District of Arizona. On August 10, Mater sunk the shaft 10 feet for Smith to show that he had improved the mine, thereby extending his lease until December 1902. On the surface, the mine was running $5 per ounce in gold, $10 per ounce in silver, and 35 percent to 40 percent lead. Both men were enthusiastic about the mine's potential. At about the same time, Smith and H. E. Casey sold the Carisa, a copper mine in Utah, for $650,000, a profit of $500,000 in less than two years. Eben and Frank also sold their interest in the Denver Mine and Smelter Supply Company to their partners the Cary brothers for $467,000, a considerable profit from their original $30,000 investment.[7]

The end of summer found the Smiths back in Denver, where Eben and Emily finalized last-minute arrangements for their move to Los Angeles, including shutting down his office in the Equitable Building. In October, the Smiths left Denver for Los Angeles. Eben opened a temporary office at the California Club in Los Angeles before renting Room 611 in the Laughlin Building, and he hired a new secretary, Charles L. Lapham. In mid-December, the Smiths completed one last piece of Colorado business when they transferred ownership of their Logan Street house, "so long one of the architectural attractions of Capitol Hill pointed out to visitors as the home of the distinguished financier," to their daughter Cora. The Carnahans moved into the house, which was much larger than their High Street house, but Charles continued to split his time between Denver and Leadville, where he remained in charge of the Resurrection.[8]

Away from Denver, the Smiths relied on their children to keep them informed about news from home, especially news about their grandchildren, which now included the Carnahans' third child, Emily Elizabeth, born on March 31, 1902. Charles Carnahan was especially good about including family news in his regular reports to Eben on the Resurrection. Near the end of February 1902, Charles wrote that the children were very anxious to visit their grandparents in California. According to her father, all Doris could talk about was swimming and how much she looked forward to it. A week later he reported that he had seen Eben's brother Mark, who was not looking well, which prompted Charles to write that he could not understand why Mark did not get out of Leadville. He also informed Eben that Lemuel was doing well in his work at the mine.[9]

Far from retiring, Smith immediately jumped back into business almost as soon as he arrived in Los Angeles. As he had done with all of his most recent mining ventures, Smith began selling stock in the Pride of the West shortly after his move. He hired John S. Cary, his former partner at the Denver Mine and Smelter Supply Company, to handle the sales. At the beginning of November, Cary wrote to Smith that sales were good and that he was doing everything he could to see to it that all the stock sold. That same day, Eben Smith and Sons of Denver leased the Garfield Mine near Crestone, Colorado, from Hugh Haukifer. A. Tipton discovered the Garfield in 1884 and sold it to Haukifer in 1885. Eben immediately built an engine room, a blacksmith shop, an elevated tramway to the ore bins, new rails, and other "general improvements." Near the surface, the vein was running from $60 to $120 in gold, and observers believed further development would reveal a large body of high-grade ore.[10]

Two weeks after leasing the Garfield, Smith remembered his retirement pledge and had attorney W. D. Reid file papers to dissolve the Buell Gold Mining Company, the property he had been managing in Black Hawk for Bela Buell since 1895. A week later, he wrote to his attorney Tyson Dines, in response to a business proposition, that he would definitely be away from Colorado most of the time in years to come and could not "afford to scatter my business all over the face of the earth." Just a few months later, though, in March 1902, the next good mining scheme came when Smith and W. W. Swetland paid $22,000 to N. S. and Emma Berray and Claude and Adele Batailleur for the Brewery and Lead Home mining claims in Gila, Arizona. A little over two weeks later, the two men paid $42,000 for a one-year lease on two copper claims on the Middle Pinto Creek in the Globe district of Arizona. These mines do not appear to have been as successful for Eben Smith as his ownership of the Carisa was, but they were still profitable. Smith's supposed retirement did serve as a good excuse to get rid of those who were trying to sell him mines if he was not interested. On June 23, Frank Sutherland wrote to Smith about a man named Garrison who wanted to sell him a mine. Sutherland told Garrison that Smith had given up mining altogether and would not be interested in the offer.[11]

Eben Smith continued his efforts to find work for his family at any of his remaining mines if they expressed interest. In April 1902, he wrote to Carey W. Thompson, then in charge of the Victor Junior, that his nephew Roy, son of his brother Henry, was looking for a job. Roy found work at the Victor Junior especially hard, as Thompson informed Smith in early July. Smith wrote back that he had made it clear to Roy "that he was not working for me, that he was working for you," and that if he wanted to change jobs, Thompson was the man with whom to speak. He went on to say that he did not want Roy "to think for a moment that I can make his paths any smoother. I want him to work out his own situation." In addition to writing about Roy, Eben also informed Thompson that Lemuel, then working at the Resurrection, was doing well in his battle with alcohol. Thompson was especially pleased to hear this news, as he knew that Lemuel wanted to "break away" from it. He wrote to Eben that "the habit is a disease," that it took a "strong fight and I hope Lem will win."[12]

In May, Smith hired Charles Mater, who worked for him at the Crown Point Mine, to prospect for him in California and adjoining states at $3 a day. On June 10, he wrote to the always cost-conscious Smith, in explanation of his expenses, that it was costly to prospect in the desert because it was necessary to bring water and food for the horses with him. For the next four years, Mater would alternate between working at the Crown Point Mine and searching the deserts of California and surrounding states for mines to present to Smith. Shortly after he hired Mater, Eben and Emily left Los Angeles to spend the summer at Palmer Lake. Frank had established a new office at 1740 Broadway in Denver, and when Eben was in Palmer Lake, it served as his office as well. He would take the train to Denver each morning and then fall into the comfortable routine he had followed for many years. After spending the morning working, he would go to the Denver Club for a roast beef sandwich and coffee for lunch and then back to the office for the afternoon before taking the evening train back to Palmer Lake. Despite appearing busy, Smith wrote to his old friend J. P. McCoy in mid-June that he was not as he once had been "but a good deal lazier."[13]

Although Smith had clearly not given up mining altogether, in mid-July he took the time to sell one of his older mining properties. The Granite had never performed as well as his other Cripple Creek mines, and it had caused him a fair number of problems, so it was with relief that on July 11 he completed the deal and resigned as both president and—along with his son Frank, D. L. Webb, Robert Reid, and William P. Dunham—a director. For his trouble, Smith received $123,353 for the 127,155 shares of stock he owned in the company. Emily Smith, her sister Sarah Rundel, and Cora Carnahan owned 5,000 shares of stock each, for which they each received $4,863. That same month, Smith and Charles Carnahan began another round of improvements at the Resurrection Mine. The Number One shaft at the mine was to a depth of 1,200 feet, at which point the miners ran into large bodies of low-grade ores. They determined that they needed to treat this ore at the mine before sending it to the smelters, so the following January, Smith and Carnahan began construction on what was to be the largest milling plant in Leadville. At the plant, the ore was roasted and then run over tables where magnetic separation removed the zinc in it. This process resulted in a high-grade zinc product in addition to iron and lead products, all of which went to the smelters. The *Denver Times* labeled the mill an "important mining enterprise, the result of which will be a bigger output from the Leadville district." The owners of the Resurrection certainly hoped the mill would also result in a bigger profit for them.[14]

By 1902, it appeared as if Smith's Colorado mines were making a comeback. The 1901–1902 annual report of the Independence Consolidated Gold Mining Company, of which Smith was still president, told the story of a mine that was in better shape than it had been for a long while. During the year ending June 30, miners accomplished a total of 2,745 feet of sinking, crosscutting, drifting, and level work at the mine—sinking the main shaft itself an additional 203 feet, to a total depth of 1,082 feet. Settling of machinery at the mine caused about $2,000 in damage to the property and a temporary work stoppage, but William Dunham reported that the damage was not serious and work was again under way. Up until the time of the shutdown, the mine had been paying $6,000 a month in royalties. Overall, Dunham thought the property was in decent shape, which was good news for a mine Smith had once wanted to shut down if it did not start paying its way.[15]

Conspicuously absent from Smith's activities throughout this time were Smith-Moffat's two largest Colorado holdings, the Victor and the Venture Gold Mining Company. The Victor had reopened the previous October, but its 1902 output was nothing compared to its heyday. The biggest problem was that the vein was still playing hide and seek with workers. Smith still thought the vein had crossed onto the neighboring Isabella property, so between August 20 and August 26, 1902, he bought 50,000 shares of stock in the Isabella Gold Mining Company through the Quentin Investment Company. He knew the vein had been a rich one for the Victor in the past and thought that if the Isabella was going to benefit from it, he might as well do the same. He did sell 1,000 shares of the Isabella stock on October 1, but he kept the remaining 49,000 shares until his death.[16]

In the ten months since he left Denver, Smith had purchased or leased a number of new mining properties and improved conditions at some of the older ones. It came as no surprise, then, when the *Denver Times* announced on August 21, 1902, that Eben was negotiating for a piece of land at York Street and Eighteenth Avenue in Denver. According to the *Times*, Smith had grown tired of the California climate after being there less than a year and had taken interest in the "new projects afoot in the state," although the newspaper never stated what these projects were. Eben planned to spend "not less than" $75,000 for the new house he was going to build on the lot. In comparison, the house Frank was building on the opposite corner of Eighteenth Avenue was reportedly costing $50,000, and when Eben deeded his old house

FIGURE 7.3. Frank and Josephine Smith's house at Eighteenth and York in Denver, ca. 1905. Denver Public Library, Western History Collection, X-27058.

on Logan Street to Cora the year before, its estimated value was $40,000. If nothing else, Eben was definitely planning to move up in the world.[17]

Eben may have tired of the California climate, but Emily had not. She quickly put a stop to any permanent return to Denver, and the Smiths were back in California in late August 1902. By then, daily offers of business opportunities had replaced the almost daily offers of mines that once plagued Eben Smith. On September 2, he declined J. P. McCoy's offer to invest in a life insurance company in China, saying it would be one of his "remotest" ideas to invest in a place so far from "God's country." He went on to say, "I would as soon think of investing in an ice plant in hell, where I would not expect to get any return until I was located there permanently."[18]

At the time, Smith had, in fact, just bought 30,000 shares of stock in a Utah ice plant. His son Frank also got involved, buying 12,500 shares. At the first organizational meeting of the Utah Ice and Cold Storage Company, the stockholders chose Eben to serve as a director for the first year of business, but for the most part, he left the day-to-day operations to others. He may not have been retired, but at the same time, he did not then feel the need to be as involved with the business as he would have a few years earlier. Mining was still Smith's first love, though, and in October, Smith hired Henry Gillespie to prospect for mines in Dutch Guiana in South Africa. This was a huge leap for a man who for years had objected to mining outside Colorado. He never invested in any properties Gillespie may have found, but the fact that he even thought of looking at mines so far away confirmed his great love for mining.[19]

In general, Smith's business life was doing well, but trouble was again brewing. William Dunham had been involved with Smith, to varying degrees, in the mining business since 1898. Their biggest venture together was the Independence Consolidated Gold Mining Company in Cripple Creek. While Smith was in Denver, Dunham hovered on the edge of his business activities, but that changed when he moved to Los Angeles. When Smith shut down his Denver office, Robert Reid's employment ceased, which meant he also was no longer Smith's business representative in Denver. Dunham was ready and willing to take Reid's place, even if not invited to do so by Eben Smith himself.[20]

The trouble started in late October or early November when Smith sent Dunham a large block of his shares in the Independence Consolidated Gold Mining Company, which Dunham was supposed to send to the International Trust Company to sell. Instead, Dunham, who was general manager of Independence Consolidated, kept the shares. Smith prodded Dunham to send the shares to International Trust, but Dunham steadfastly refused. On November 21, he wrote to Smith that it hurt him to have Smith continually refer to the matter "as though you thought this stock and this whole transaction were not as safe in my hands, as far as your interests are concerned, as they would be in the hands of any trust company in the world." Dunham's stubborn refusal to part with Smith's Independence stock was the first in a long list of troubles he caused.[21]

At the same time he was holding on to Smith's Independence stock, Dunham was also causing problems at the Victor. Dunham was part owner of the Thomas Company, which engaged in mining next to the Victor property. David Moffat claimed, in all probability with good reason, that Dunham and the other owners of the Thomas Company had taken ore out of the Victor Mine. Moffat wanted the land owned by the Thomas Company deeded to the Victor, and he threatened to sue, if necessary, to accomplish this. Smith's attorney, Tyson Dines, investigated the matter and found that the Thomas miners may not have encroached on the Victor property. According to Dines, the patented description of the Victor property did not include the very small piece of land in question, but the original survey did. Dines said the outcome of the dispute might hinge on which property description the courts accepted as valid.[22]

In the same letter in which Dunham told Smith that his Independence stock was safe with him, he also told Smith that he would deed the Thomas property to the Victor. He must have hoped this would smooth things over, but it did not, as Dunham neglected to actually deed the property over and Moffat continued to pursue the issue. On January 9, 1903, Dines wrote to Smith that Dunham had still not followed through on his promise, leading to increased frustration on the part of Smith-Moffat. It was not until July 8, 1903, that Dunham finally deeded the land to the Victor, but even then, he delayed signing the final papers for as long as he could. By then the trouble between Smith and Dunham had gone from bad to worse.[23]

On December 20, 1902, Eben received one of the only harsh letters David Moffat ever wrote to him. Smith and Moffat had become the owners of the Jack Pot group of mines in Cripple Creek in the early 1900s. In early December, Dunham signed Smith's name, as his agent, to the escrow papers of the Jack Pot group, leasing the mine to Dunham and a man named Woods. In addition to getting control of the mine, Dunham received a $20,000 commission for leasing the mine to himself. Moffat was outraged at Dunham's trickery. He wrote to Smith that at the time Dunham had signed Smith's name to the papers, which he may have done without permission, he and Woods had already taken $300,000 worth of ore out of the mine. The option Dunham gave himself did not expire until August 1903, and by then Moffat believed Dunham and Woods would have "completely gutted" the mine, leaving Smith-Moffat with "the 'hole' . . . and no money." After warning Smith to be careful about such men, Moffat went on to encourage him to get Charles Carnahan or somebody else to look into what had taken place. Moffat had already arranged to have Tyson Dines look into the matter in an effort to get the agreement overturned on charges of fraud. He then told Smith that he wished he were in Denver "to help dig out this fellow Dunham. He is a Daisy," adding that he thought 1903 would be a "good year for thief catching." Moffat, at least, was out for blood and saw Dunham as the deserving target.[24]

After writing to Smith, Moffat took the matter up with Carnahan. Moffat convinced him to write to Smith, which he did in mid-January. Carnahan told his father-in-law that up until then he had not offered an opinion on any of Eben's business affairs that he was not personally interested in, but since Cora was his wife, he felt it was his duty to "speak for her, or in the same manner exactly as if I was your own son." Carnahan reminded Eben that before moving to California, he had told his son-in-law that he believed Dunham and Reid had used their association with him to make money for themselves, to his "loss and detriment." Taking this statement into consideration, Carnahan met with Dines, who agreed that the two men's behavior was improper and that they had proven themselves unworthy of trust. Carnahan and Dines also believed Dunham and Reid still had large interests of Eben's under their control, which could lead to "severe loss as well as a great deal of worry and distress to you." How dishonest Reid may have been is open to question. Smith and Moffat did not mention him as a problem as this issue was unfolding, and in Reid's defense, he had little contact with Smith after his move to Los Angeles.[25]

Tyson Dines wrote to Eben that he and Carnahan had discussed the matter with Frank, and he thought Eben should give Frank or Carnahan power of attorney over his Colorado businesses, especially those in which he had interests with Dunham and Reid. Carnahan agreed with this plan, as he and Frank were "here on the ground, younger and more active." It did not matter who Eben chose, Carnahan said, as he and Frank agreed on the issue. If Eben saw fit to choose either one of them, Carnahan promised that the other would "be consulted in all matters of any consequence and we will act together."[26]

On January 21, Eben wrote to his son Frank that he felt "very grateful" for the suggestions he and Carnahan had made. His letter revealed that his business relationship with Dunham had been even more costly than anyone realized. In 1900, Eben had spent a great deal of money defending his interest in the Hull City Placer, a Cripple Creek mining property in which he had invested with Dunham and William Otis. The Hull City Placer ran into trouble because of alleged wrongful and extravagant expenditures by Dunham, which led the owners of the property to file suit seeking a receiver for the property despite Dunham's claim of innocence. Smith invested $200,000 of his own money to try to recover his interests. The men lost the case, and altogether the deal cost Eben $400,000. Despite the losses, he did not think it was advisable to turn the business over to his son or his son-in-law because doing so would only antagonize Dunham and make it a great deal harder to get a settlement out of him, as he was a "stubborn brute." Eben told his son that he was "not so far away from Denver that I could not go there at any time to fix this matter up."[27]

Smith did not travel to Colorado to fix the matter up, and it eventually faded into the background. William Dunham treated Smith with a disrespect that no other person ever got away with. He was certainly dishonest, and he waited until Smith was out of state and unable to take immediate action to use his position to take advantage of Smith. What is even more surprising is that Smith continued to conduct business with Dunham after this time. He did finally realize, though, that he needed a trusted person to take care of business for him in Denver, and on November 7, 1903, he gave his son Frank power of attorney to act for him in all his Colorado affairs—but it was too late. When the trouble with Dunham first began, Moffat had written to his longtime friend that "taking to your bosom these untried friends is a little dangerous, Ebenezer." In the Dunham case, that trust, placed in the wrong person, had cost him a great deal of money.[28]

Money was weighing heavily on Smith's mind by 1903. Requests for charitable assistance continued to pour in to Smith's Los Angeles office during this time, but he was feeling the pinch of supporting so many people. On September 2, he wrote to William Lessig that he could not help him, as he had "so many poor people to take care, many of them who are in no way connected to me, that I do not feel like taking on any more people at the present time." While he sympathized with Lessig's condition, he knew that would not give him "any bread." Despite turning down Lessig, Eben was still very generous toward his family. Between April and August 1902, his son Lemuel left the Resurrection for a job as an engineer at the Victor Junior in Oregon. Carey Thompson made sure to keep Eben informed of his son's progress, and near the end of August, he reported that all was going well for him. Eben wrote back that he was glad to hear that Lem was getting along alright, and he was sure that Thompson would "find him a first class man with your engines." Showing that he was very pleased with his son, he closed by asking Thompson to give Lem "my compliments," something Eben did not hand out freely. The following February, Thompson wrote to Smith that Lemuel had slipped a little in his battle with alcohol, but he did not think it would happen again. This came just three months after Thompson had proudly reported that Lemuel was "getting to be quite a family man," so his slip was sad news indeed to his father. Lemuel and his wife, Nellie, soon left Oregon and returned to Leadville in early May.[29]

Smith may have been feeling less charitable than he had in the past, but he had no problem buying stock in any companies he thought might be profitable. On November 20, 1902, he bought $5,000 worth of stock in the Murray M. Harris Organ Company of Los Angeles. Four days later Smith paid $25,000 to buy all the stock in the Pacific Wireless Telegraph Company and the Continental Wireless Telephone and Telegraph Company that belonged to A. M. Stevenson and D. C. Bailey. He immediately took steps to consolidate the two wireless companies, first under the name International Wireless Telegraph Company and then Pacific Wireless Telegraph Company.[30]

Smith reported on his new investments, which were well outside his mining expertise, to J.O.A. Carper, a friend in Denver, at the beginning of 1903. He had known Carper for some time and had invested money in a milk plant Carper was trying to operate. Smith had also given Carper permission to try to sell Estamere in January 1903, as Eben did not expect to be using the house in the future. In February, Smith wrote to his friend that he had gone into a "good many one horse things in this country [Los Angeles] and some four horse affairs, some of which look very promising, notably the Wireless Telegraphy."[31]

There was a noticeable change in Eben Smith's business affairs in late 1901 and 1902. The Colorado mines were under the direction of Moffat, and with the exception of the Resurrection, he received no reports on them. Freed from having to show a profit to stockholders, Smith could search for mines in California and South Africa, dabble in copper mines in Arizona, and play the stock market. He hoped to add to his fortune with these investments, but in large part it was the thrill of the game that excited him. He obviously enjoyed himself, which was good because his one-horse and four-horse investments were about to become a major part of his life.

8

"I Feel I Am in Hell's Hole"

Mining matters continued to occupy some of Smith's attention as 1903 began, although not as strongly as some people would have liked. In May, prospector Charles Mater was back at the Crown Point Mine for one of his periodic stops there and was disappointed that Smith had not visited recently. He wrote to him on May 4 that he "was in hopes to see you out here before now," as he was very anxious to show the boss the mine of which he was so proud. While certainly very interested in the Crown Point property, reports from Mexico had a greater hold on Smith's attention. The previous September he wrote to Mater that he thought Mexico was about to become the new hot spot for mining. In fact, Smith wrote, he thought it was "already there," and he accordingly shifted his mining focus to the area. He quickly found a mine whose prospects pleased him, the Pinos Altos, and in June he wrote to both the Central Trust Company of New York and Tyson Dines in an effort to raise money for the venture. Within a month, as he sought more investors, he again sounded like those who had written to him in years gone by offering him mines for sale. Writing to Robert Evans, he said the Pinos Altos property

DOI: 10.5876/9781646421794.c008

was "the greatest thing I have ever seen in my experience in mining," and he wrote to John Farish, "I cannot say I have ever seen any other of such magnitude." He proclaimed to the investment company Hornblower and Weeks that the mine would "produce 1,000 tons of ore a day for a very great number of years."[1]

What was even more amazing about the property, which included the Esperanza and Dolores Mines in addition to the Pinos Altos, was that William Dunham was the man who had gotten Smith interested in it. In early August, when progress at the mine slowed to a crawl and Smith's enthusiasm for it began to diminish, Dunham once again went on the attack, writing to Smith that he was "overly critical" and that he should appreciate the difficulties such a deal was sure to encounter when the market collapsed, as it had done. He told Smith that his letters, which he was always glad to receive as long as they were not critical of him, should contain more encouragement and less fault-finding, which was "unwarranted." Unmoved, Smith withdrew from the Pinos Altos deal on August 18. He still thought the mine was a good one, but he told Dunham, "I have discovered long since that one cannot indulge in all the luxuries of this world without means to procure them."[2]

The situation at the Murry M. Harris Organ Company was equally problematic. On July 16, Smith wrote to Dunham that the organ company was in serious financial trouble. He and the other investors had discovered that Murray Harris was "pretty crooked" and that the organ company was "badly involved" in his financial dealings. With the company about to go bankrupt, the stockholders realized they would be personally responsible for its debts. Smith's $25,000 investment made him the largest shareholder and would have made him responsible for one-quarter of that debt. He quickly realized that he was going to be "obliged to bolster that thing up in order to put it on its feet."[3]

In early August, Smith wrote to his former secretary, Robert Reid, that he had loaned Harris $10,000 to try to save the organ company and that "I feel I am in Hell's hole." Harris's efforts to save his company had little impact, and Smith unofficially took charge of the company and started paying off its debts. At the August 12 stockholders meeting, the other investors selected Smith, the largest stockholder, as the company's new president and also chose a new board of directors, excluding Murray Harris. Harris's questionable finance methods led some to question whether they should even allow

FIGURE 8.1. Eben Smith, early 1900s. Louis O'Kane Collection.

him to remain an employee of his own company. Reid informed his old boss that he was "heartily sorry" for the financial difficulties he had encountered with the organ company.[4]

Smith's other new venture, the Pacific Wireless Telegraph Company, aimed to connect the Pacific Coast of the United States with the "Orient" and Hawaii through telegraph wires. Smith hired General Albert L. New as general manager of the company, but he and New had differences of opinion on almost every front. In particular, New wanted a monthly salary. Smith argued that a salary was impossible as it would have to come out of his own pocket, and on July 24, he wrote to New that the two should put aside their differences and "consider that we are all working for the same ultimate purpose—success." For the time being New agreed, but the truce would not last long.[5]

While Smith's business troubles mounted, there was also trouble at home. Frank and Josephine Smith and their three children arrived in Los Angeles for an extended visit with Eben and Emily in April 1903. The Smiths hired a nanny, Alida Buelow, to help care for the children. On the afternoon of April 22, smoke began pouring out of a locked closet in Frank and Josephine's bedroom at the house. Servants broke open the door to discover a fire, which was luckily contained because the closet had been closed. Josephine was not in the house at the time, but when she returned and heard about the fire, she immediately went to the closet to look for some jewelry she had hidden in a compartment there. While most of the jewelry in the closet was untouched, along with two watches on a dresser in the bedroom, nine rings valued at about $2,000 were missing. Los Angeles police detectives soon arrested Buelow for the crime. Buelow had a good reputation as a nanny, but according to the police, she had recently gotten involved with a less than reputable man in San Francisco and stolen the rings to help finance her relationship with him. The police found the rings, along with $15 in gold and a $10 greenback, in the tank of one of the toilets in the house, where Buelow put them after first burying the rings in the yard.[6]

On August 9, even more shocking news came from Colorado. The day before, while attending a horse race in Denver, Charles Carnahan became convinced that a man named Hugh Swearingen had insulted Cora. The two men fought at the race, resulting in a bad beating for Carnahan. Angry about this, Charles tracked Swearingen to the office of George Vallery, general agent for the Burlington Railroad. The staff at the Burlington office thought nothing of it when Carnahan walked through the door or even when he opened the gate and walked past them to Vallery's office. Without saying a word, Carnahan took out a gun and fired point blank at Swearingen. Vallery ran out the door to get help, and Charles fired three more shots at his target; all four missed. The two men then fought, and Carnahan hit Swearingen repeatedly with the gun. One final blow knocked Swearingen to the ground. The police who responded to Vallery's pleas for help arrested Carnahan, but within a few hours he was free on the verbal bonds of Smith-Moffat attorney Charles Hughes and David Moffat himself. A weary Smith reported the incident to Dunham and said his son-in-law had fired the shots from a distance of "probably not more than six or eight feet" from Swearingen.[7]

Angered by the shooting, Emily Smith wrote to Carnahan's mother, Elizabeth, that he had "never contributed anything to the support of [his] family." A shocked Charles told Eben that he could prove he had spent well over $50,000 on his family's care. As the family waited to see what would

FIGURE 8.2. Emily Smith, 1909. Michel Tritt Collection.

happen, on August 22 the *Denver Post* reported that public sympathy was with Cora and that prosecution of the case might turn out to be very bitter, as "Mrs. Carnahan's . . . father is not apt to let the matter drop if his daughter insists upon a vindication of her name." The grand jury met in Denver on September 5 and refused to indict Charles Carnahan, but many thought he would still have to stand trial on assault charges. Three weeks later, on September 26, the newspapers reported that he was set to plead, but according to the *Los Angeles Herald*, the case was dropped when Swearingen refused to appear in court. In the months after the shooting, rumors circulated that Charles was going to divorce Cora to spare her further embarrassment, but in March 1904, Eben traveled to Denver to negotiate a reunion between his daughter and son-in-law.[8]

Business would normally have been Smith's escape from such troubles, but there was only more trouble to be found there. William Ramsay, an employee of the Murray M. Harris Organ Company, had been investigating the financial practices of Murray Harris and the bookkeeper, Mr. Dewey. On August 27, he reported to Smith that he found nothing "to note criminal intent or wrongdoing" on the part of the two men outside of a large block of stock Harris had issued without approval, but he found the "greatest extravagance and carelessness in matters of expense" and accused Dewey of incompetence for the numerous bookkeeping errors he made. Ramsay's investigation only confirmed the serious financial problems the company, and Smith, faced. It was becoming increasingly clear to the new board of directors that they needed to get rid of Murray Harris. Ramsay was doubtful that Harris, whom he considered "thoroughly dishonest," would be able to adapt to the new circumstances he found himself in at the company, as he was working only to "straighten out his affairs." In addition to Harris's "wrong-doing," a contract Harris signed for a particular organ was seriously draining the company's resources. Employees referred to the organ in question, destined for the 1904 World's Fair in St. Louis, as the "big organ" or the "big instrument" because, if completed, it would be the world's largest pipe organ. Harris had agreed to build the organ at a ridiculously low price that would not cover constructions costs, and Ramsay told Smith the contract was a financial mistake and a "tremendous over-reaching by Mr. Harris of all reasonable business conditions."[9]

Smith agreed with Ramsay's assessment of Harris, writing in early September that Harris was "no good on earth. He is full of badness, and has no good qualities, and is a[n] unsafe man to be connected with us in any way." Smith and Ramsay wanted Harris out, and they planned do whatever they could to rid the company of him. In addition, they decided that it "would be policy" to drop the Murray Harris name from the company if they wanted to have any chance at saving it. The two discussed possibilities for a new name but could do nothing until Harris was out of the picture. Ramsay moved quickly on that front, proudly announcing to Smith on September 19 that Murray Harris no longer worked for the company, but he did not say how he had accomplished that. Getting rid of Harris turned out to be easier than renaming the company, as three months later they were still trying to come up with something more suitable. The directors narrowed the choice down to two possibilities—the American Art Organ Company and the Los Angeles Art Organ Company—before finally deciding on the second. In early February 1904, Smith wrote to Ramsay that he had secretly hoped for American Art because of the more national reputation it would have given the company, but he lived with the change as it was.[10]

As the debate over the new name for the organ company carried on in late 1903, work at the Pacific Wireless Telegraph Company continued to charge forward. On November 1, General New was able to report to Smith that the system was "working perfectly" between Port Townsend and Fort Casey, Washington. The company had also received permission from the US government to install telegraph stations on revenue cutters, further increasing their reach. New went on to report that workers were finishing installation of a station on the Revenue Cutter Grant and that stations at Friday Harbor on San Juan Island, Washington, had successfully received several test transmissions. General New assured Smith that he could not lose money.[11]

In spite of New's assurances, the Pacific Wireless Telegraph Company and the Los Angeles Art Organ Company continued to drain Smith financially into 1904. Just four days into the new year, Smith wrote a check to Pacific Wireless for $2,500. Six days later, Smith learned some disturbing information about one of his employees at Pacific Wireless that put the employee on a par with William Dunham. Smith informed General New that Armstrong, one of the men selling stock in the company, had been doing so without turning in the money from those sales. In his letter to New describing what had happened, Smith wrote, "I am rather unfortunate in running up against crooked men." The situation was no better at the organ company. In January, Eben and Emily sold a piece of land they owned at Seventh and Alameda in Los Angeles to the organ company for $10. The company built a new factory on this land, the expense of which was borne by Eben, but it did not help. By the end of February, he was so frustrated with the situation there that he wrote to William Ramsay that if not for the big organ, he would "drop the whole business now and take my loss." The company had been working on the St. Louis World's Fair organ for six months at that point, and so far, it had cost $60,235, much to the shock of Smith. Contracts for other, smaller organs in no way came close to covering the $67,000 total it would eventually cost to build the big one. Despite Smith's frustration with the company, the big organ attracted a great deal of attention.[12]

An article in the April 23, 1904, edition of *Scientific American* gave a detailed description of the organ, which had 10,000 pipes and 140 stops and weighed 125 tons. When the organ was finally completed, its weight caused further problems for the company, as it cost a huge amount of money to ship it from Los Angeles to St. Louis in ten freight cars. The organ measured 100 feet by 27½ feet, was 40 feet high, and required a total of 80,000 feet of lumber for construction. The console alone required 10,000 feet of quarter-sawn white oak. The largest pipe was 32 feet long, and in a publicity photo, a Shetland pony posed inside that pipe. The article also demonstrated why the organ was such a blow to the company's financial affairs. The company employed a great number of men in constructing the components, in addition to placing the 1,300 magnets, 115 miles of wire, 90 pistons, and many other parts that made the organ actually work. The company also hired a number of women and girls to sew together the hundreds of sheepskins used to cover the pneumatics inside the organ's 12-foot-long chests. While the cost of the organ kept the company from being profitable, Smith hoped that successfully completing it would bring more business.[13]

During April and May, Smith loaned the organ company an additional $15,000, making it clear that the only thing keeping the Los Angeles Art Organ Company and Pacific Wireless Telegraph afloat was Eben Smith, much to his dismay. In early March 1904, he wrote to General New that he was tired of paying out money for the telegraph company and that it was long past time to get it on a paying basis. In the same letter, Smith revealed that trouble had developed at the organ company between him and Ramsay. He told New he was afraid of Ramsay, who was handling a very large amount of money for him, and that he could not "afford to have a man who is attending to my business on the one hand and fighting it on the other." Smith did not elaborate on what he meant by this, and there is no further indication of what the trouble may have been, but it looked as if Smith found himself once again trusting a man who did not deserve his trust.[14]

By then, the organ company had opened an office and showroom in New York City under the direction of Arthur Scott Brook. Brook tried to avoid spending too much money on decorating the showroom by "exercising economy with a good general appearance," and he sent letters that went over his expenses and thoughts for the company down to the smallest detail, much to Smith's satisfaction. Just as Smith had expected reports from those who were in charge of his mines in the 1890s, he expected the same kind of reports from those who were helping him make music and string wires in the early 1900s. He was not afraid to let his employees know he would replace them if they did not achieve the results he desired, though more often than not, nothing seemed to come of the threats.[15]

The organ company added "Builders of the Largest Organ in the World" to its letterhead in 1904, but regrettably, as far as Smith was concerned, this newfound fame did not translate into financial gain. In July, he loaned it $5,000, followed by $3,000 in August; a statement dated September 2 revealed that the organ company owed him $156,847. In October, that total rose yet again when he loaned the company another $7,500. By then, he might have been wishing he had settled for one-quarter of the company's debts and let it go bankrupt. However, Smith's success in difficult mining ventures led him to believe he could turn around a failing organ company. He was not one to give up easily, and he was far from finished with the organ company.[16]

The situation at Pacific Wireless was only slightly better. Little happened there between the time in March, when he told General New it was time to put the company on a paying basis, and October, when he loaned it another $1,000. By January, that company owed him $96,000. Smith, who in 1901 thought he was worth only $1 million, in two years loaned a quarter of that money to two companies with no guarantee that he would see it again.[17]

By late 1904, Smith had grown very frustrated, and it took its toll on the seventy-two-year-old's health. On September 4, he suffered what the *Denver Post* described as a paralyzing stroke at his home in Los Angeles. Friends in Colorado received the news by telegram and feared that because of his age, there might be a "fatal termination of the attack." Employees wrote to encourage Smith in his recovery, including Arthur Scott Brook who said his father had suffered a similar stroke and fully recovered. His old mining associate Charles Limberg also wrote to say that he hoped Smith had fully recovered. The fears of his Colorado friends proved to be groundless and the severity of the stroke exaggerated, as Smith fully recovered and was back at work by September 24, loaning yet more money to the organ and telegraph companies.[18]

While most of Smith's attention was on organs and telegraphs, he continued his mining activities on a limited basis. A more violent labor strike rocked Cripple Creek starting in 1903, where Smith was still part owner of the Victor, Venture, Roxana, and other mining claims. However, with production at the mines almost nonexistent, he never commented on this latest round of labor trouble and did not reply to the one letter Moffat sent him about the strike. Not even the fact that former Victor superintendent Sherman Bell was in charge of the Colorado militia during the strike brought a comment from Smith. Eben had also gone in on another mining venture with his brother-in-law C. W. Denison when the two leased the Silver Lake Mines in Silverton. Denison was in charge of work at the mine, and Eben's son Frank handled most of the business side for him. Charles Mater had given up prospecting and was back at the Crown Point Mine in Arizona, still hopeful that it would prove profitable for him and Smith.[19]

Unlike the majority of his Colorado mines, the Victor Junior in Oregon continued to be a productive property. Carey Thompson reported to Smith on January 7, 1904, that the mine was shipping several thousand dollars' worth of ore every month, and the December 1903 cleanup of the mine, for example, had yielded $14,000 in gold bullion. His nephew Roy continued to work at this mine, although he seemed never to get over how hard the work was. Thompson reported to Eben on April 19 that Roy had left, but, unable to find anything better, he was back by May 4. Eben heard little from Lemuel, who had gone back to Leadville the previous year, but in late 1904, Eben started sending him $25 a month. Smith continued to buy new mining properties as well, including that of the Joseph Gibbons Consolidated Mining and Milling Company in San Juan County for $112,500 in July 1904.[20]

By late 1904, Eben was anxious to escape from his business concerns for a short while. After moving his Los Angeles office from the Laughlin Building to the sixth floor of the Braly Building, he and Emily left town in early November. After a quick stop in Denver, they made their way to St. Louis for a visit to the World's Fair. There, Eben saw the big organ in action, and he wrote to his new secretary, Charles Lapham, "I assure you the great organ is all and more than they claim it is." From there it was on to the Smiths' beloved New York City for a much-needed rest, but his enormous outlays of money to the organ company and the wireless company were weighing heavily on Eben's mind as 1905 rolled around. Much to Charles

Mater's dissatisfaction, he received a letter from Smith dated January 8, in which Smith said he did not want to put any more money into prospecting and even mentioned the possibility of selling the Crown Point and the more recently purchased Golden Fleece Mine.[21]

Smith became so frustrated with Pacific Wireless at the end of January that he told General New that at times he felt like "throwing up the whole business and letting it go to the dogs." He realized he would lose a large sum of money by doing so, but whether such an outcome would be preferable to putting more money into the venture was a "serious question" in his mind. This came after New again requested a salary of $300 a month, which Smith again denied because it would have had to come out of his own pocket. The frustration continued to grow when he sent another $400 to New in mid-February for the construction of a wireless station in Victoria, British Columbia. He had already put $1,800 into this particular phase of construction, and he wanted it operational. The antagonism between New and Smith continued to worsen, especially when New wrote that the company might as well be called "Mr. Smith's Wireless Telegraphy" since he had control of the stock and the board of directors. Smith blew up, writing New that it was a lot of "damned rot" to say such a thing to him. He went on to state that if he "could get out what money I have in the concern I would be glad to turn it over to you or anybody else." The situation got so bad that when New later threatened to resign, Smith told him he would gladly accept his resignation if he wanted to give it.[22]

The situation between New and Smith sometimes spilled over into his relationship with other employees, often over trivial matters. S. L. Phillips, another employee of Pacific Wireless, had been sending telegrams collect to company headquarters. To a man like Smith, who strongly advocated keeping expenses down, such a move was bad policy, but to a man already on edge it was disastrous. He berated Phillips for sending the telegrams collect, and Phillips apologized, saying he would know in the future to attend to his business and not to Smith's, as it "certainly seems to be a thankless job." He wrote that he knew Smith had more on his mind and more annoyances with the company than any ordinary man should but asked him not to "go at your friends with a club, who are working for your interest and the interest of your Company."[23]

At times, Smith felt like he was the only one working for the interest of the organ company. In February 1905, he hired F. L. Spaulding to build a one-story addition onto the new factory building, but it did not help matters much as the company faced increasingly stiff competition from the Aeolian and Hutchings-Voley Organ Companies. In June, W. B. Fleming wrote to Eben that "our company builds the best and gives the very best satisfaction," yet it was losing business to these other firms. Fleming saw the distance of Los Angeles Art Organ from the East Coast markets and the high shipping costs incurred as a result as the biggest disadvantages facing the company and planted the idea of building a factory in the New York area. Smith and others at the company jumped at the suggestion, and by the end of the year, a new factory was operational in Hoboken, New Jersey. They also changed the named to the less local-sounding Electrolian Organ Company.[24]

Smith was able to put aside his doubts about both companies when the situation called for it. In early August 1905, he received a letter from Fannie L. Lash, who was interested in buying stock in the telegraph company. He was pleased to inform her that the company was busily erecting what would be the largest wireless station in the world on Mount Tamalpais outside San Francisco. As the largest stockholder in the company, Smith was "all the time investing money in the same, which I would not do if I were not a believer in the system." Writing to his old friend Henry Wolcott, who questioned why organs from Electrolian cost so much, Smith explained that the higher cost was because an organ from the Electrolian Company was of a better quality than he would find at any other manufacturer. He told Wolcott that "rather than build cheap organs for the sake of getting the market for them we would quit the business entirely."[25]

Not satisfied that he had his finger in enough pots, Eben Smith became a large investor in two more enterprises in 1905. The Bakersfield and Ventura Railway, his fourth railroad, was attempting to build a line between those two cities. Smith was president of the railroad, but as with all his other capitalist ventures at this time, his chief role was as supplier of the capital. The railroad caused its share of problems for Smith. In late June, the company decided to buy a used locomotive from the Salt Lake City Railroad, but it turned out to be in poor condition and needed a great deal of work and money to get it in working order. The road also had trouble getting dump cars shipped from the National Coal Dump Car Company, and it fell to Smith to straighten out the situation. Far from a quick moneymaker, the railroad presented one crisis after another and would continue to cause problems even after his death.[26]

He also invested in his old friend Frank Sutherland's new drill company, buying 76,752 shares in the newly created National Gas Drilling Company. Smith's role in this new business venture also marked the return of William Dunham. Work on the Panama Canal was pushing forward in 1905, and Dunham wanted the right to sell the new drills to the US government for use on the canal. Eben did not pursue getting Dunham this right, but he did get his son Frank a contract to sell the drills in five states. His father told Frank that the drills would cost $500 to make and that he could sell them for anything over that amount. As with his other business ventures during this time, production of the new drill ran into problems, and Eben became increasingly irritated with the delays Sutherland frequently encountered.[27]

Delays were just about all Smith encountered at the telegraph and organ companies as well. After more than three months of delays, the Mount Tamalpais wireless station was no farther along, and Smith was increasingly impatient about getting on with the work. At the organ company, affairs continued to decline, and Smith loaned it another $2,000 at the beginning of November. Even the mining business, already slow, managed to encounter delays. In early December, Smith hired George Kislingbury to examine and try to secure a number of possible copper mines in Sonora, Mexico. Smith also finally convinced Moffat that it was time to sell the Resurrection Mine. After the mine's sale, Charles Carnahan, with the backing of David Moffat, opened the C. T. Carnahan Manufacturing Company at 1724 Lawrence Street in Denver. The new firm's main business was drills, putting it in direct competition with Eben and Frank's investment in the National Gas Drilling Company. Carnahan reported to his father-in-law that he had seen one of his new gas drills in a store window but did not have the time to examine it closely.[28]

One other mining matter came to Eben Smith's attention at the end of 1905, and it brought him into contact with another old friend. George Ruple wrote to Smith concerning rumors he had heard that the Joe Givens property (the men were probably referring to the Joseph Gibbons property Smith had bought the year before) was for sale. Smith was happy to hear from yet another old friend and reminded Ruple that "the last time I heard from you I borrowed a shirt from you." He confirmed that the property was for sale, but it was under the management of C. W. Denison. Smith suggested that Ruple should write to Denison directly and that he could probably find him in Denver, as he thought Denison had a girl there, "and you know that brings a fellow around once in a while."[29]

Eben Smith loved business, but at the end of 1905, business seemed to bring him nothing but frustration. Smith's personal life was the one bright spot in all this frustration. Much to their pleasure, Eben and Emily's children and grandchildren were frequent visitors to their house in Los Angeles, particularly Cora and her three children. The children often stayed for extended visits with their grandparents, as Harold Carnahan had done in 1903. This was again the case in June 1904, when young Emily Carnahan was staying with her grandparents. Eben had to go to Denver on business while she was there, and, perhaps still fearful about Alida Buelow's theft and arson at the house the previous year, he sent a telegram to his wife on June 2 urging her to keep an eye on the baby, as a "party might conclude to kidnap her in our absence."[30]

Emily remained safe, but if kidnappers had come for her, her grandmother might have been the best person to get her away from them. During her days in Leadville, Emily Smith earned a reputation for having one of the fastest horses in town, and in 1905, she urged Eben to buy an even faster mode of transportation. On September 12, Eben bought Stevens Duryea Number 1206 from the Felker Automobile Company in Denver for $2,725. He had the car shipped to Los Angeles, and on October 20, Lapham reported to Frank that his mother's car had arrived the previous Wednesday "in good shape, and she is now enjoying it as the weather is exceedingly fine." Not ready to rush into the joy of motorized movement, Eben hung on to his carriage and three horses—Bro, Lan, and Vic. In 1906, the Smiths traded in the Stevens Duryea for a Packard.[31]

As 1905 ended, Eben Smith was preparing himself for another busy year ahead. Although the year had proved frustrating in many respects, he hoped 1906 would be a productive and profitable year for his businesses. He had a number of other business opportunities in the works, particularly with his continuing search for mines in Mexico and the drill company. But, as was so often the case at this time in his life, things would not go as he planned.

9

"The Dean of Mining in Colorado"

On January 2, 1906, Eben Smith wrote to Arthur Scott Brook that he knew it would take a great deal of hard work for the Los Angeles Art Organ Company to "get in where we belong," but he believed it was possible. Requests for help continued to find their way to Smith's desk, such as the one he received on January 6 from an old miner named Osborn. Addressing Smith as "the miner's friend," he told him that if he sent any money he could spare, God would bless him and Emily. Smith suggested that Osborn "call on God for any favors of this kind" and advised him to cut out luxuries like whiskey and tobacco, do a little hard work, and trust in God. Smith was particularly upset with Osborn because he had been sending letters to his house, and he informed him that "there is no reason on earth, why you, not knowing me, should call on me to help you." Smith also sold his stock in the Independence Consolidated Gold Mining Company in January for 30 cents a share.[1]

A little over three weeks later, on January 25, Emily Smith sent a telegram to Cora, warning her that her father was seriously ill after suffering another stroke. The following day Eben's secretary, Charles Lapham, sent a

DOI: 10.5876/9781646421794.c009

FIGURE 9.1. Frank Smith's caricature from Representative Men of the West in Caricature (1904). Author's collection.

telegram to Frank, asking him to come to Los Angeles to help attend to his father's business. Frank was already heavily involved with the Bakersfield and Ventura Railway and the Pacific Wireless Company and was so well-known that he was included in the 1904 book *Representative Men of the West in Caricature*. By mid-February 1906, Frank was in Los Angeles handling all of his father's business interests. The same day Lapham asked Frank to come to Los Angeles, George Kislingbury withdrew from the mine deal in Sonora, Mexico, after problems developed with the Mexican government. It was the last major mine venture in which Eben Smith was involved.[2]

As time passed, it became clear that Eben was not recovering as quickly as he and his family had hoped he would. In early February, Lapham sent a telegram to Brook in New York informing him that Smith was sick and still not attending to business. Nine days later, Lapham wrote to F. R. Coffin, another representative of the organ company, that Smith was doing better but would almost certainly stay away from business through the summer. His illness did not keep the organ company from needing his financial assistance, however, and between March 5 and April 18, he loaned it a total of $9,500. Eben did manage to attend to some minor business during this time, but for the most part Frank and Lapham were in charge.[3]

In early March, Smith found himself drawn into a situation that had developed at the Crown Point Mine. Charles Mater was back there full-time, and it became clear that he had invested a lot of himself in the property. On March 8, Charles Mater Jr. wrote to Smith from Arizona that his father had gone "insane" and was refusing to leave the mine because he was "under a deep impression that someone is trying to defraud him out of his interests." The younger Mater wanted Smith to arrange a bogus sale of the mine so he could get his father away and get him the treatment he needed.[4]

Smith answered on March 10, and within a week the two Maters left for California. While traveling, Charles Mater wrote a letter to Smith explaining what had happened at the mine. The trouble started when someone delivered a load of 2-inch lumber to the Golden Hill mill without Mater's knowledge, and he promptly ordered the person who brought it to take it away. Mater suspected that a man named Thomas A. Irving delivered the lumber as part of a plot to take over the mill. Irving had permission from another miner to take over his claims in the Crescent District, but no one passed that news on to Mater. Irving also wanted to do some work on the Golden Fleece Mine, telling Mater "it shall cost [neither] you nor Eben Smith anything." Mater agreed to this with the conditions that he be in charge of the work and that no monuments (survey markers) on the Golden Hill mill be disturbed, which became another sticking point as Irving thought some of the monuments had to be moved.[5]

In Mater's narrative of the story, Irving then tried to "make friends" with him over a drink in front of Mater's cabin, but Mater objected, telling Irving "I make no friends over a bottle [of] whiskey." Undeterred, Irving demanded a place on the Crown Point Mine's water right. When Mater refused, Irving came into his cabin with a gun in his hand. Mater drew his gun in response, and Irving "fell out backwards" from the cabin. Even though Mater wanted to shoot Irving, he let him go. It was at this point that Mater became convinced that someone was trying to steal the Crown Point Mine and his friends alerted his son. Five months later Smith received a telegram from Mater, who informed him that he believed their mining claims to be better than he had ever thought and that he still intended to work them. It was the last Smith heard from Mater.[6]

The situation at the Pacific Wireless Company was no less tense. Taking advantage of Smith's absence from the office, General Albert New demanded that Lapham give his son Frank a job with the company at a salary of $75 a month. Smith personally handled the situation, and after chastising New for taking such action, he stated his absolute refusal to hire Frank. He also told New that he did not want his son hanging around the office as a "spy." Smith was disgusted by New's "antics" and reminded him that he had once before said he would accept his resignation. He now said that if New's bad behavior kept up, he would demand his resignation.[7]

At the organ company, the disgraced Murray Harris also made his reappearance. According to Arthur Scott Brook, W. B. Fleming priced the organs the company produced too high and was rumored to have taken and destroyed plans for organs when leaving the factory at night. Brook told Smith that what Fleming was doing amounted to sabotage, as Harris was back and trying to reunite all his former employees, including Fleming, in a new company. Brook wrote that the organ company was faltering because Fleming was the "canker worm eating the life out of the business." Near the end of May, Smith was feeling well enough to attend to a little business at the organ company on his own. Still trying to sell the big organ, he wrote to beer baron Adolph Busch trying to interest him in it for the new auditorium under construction in St. Louis, and he replaced Brook with F. R. Coffin as the company's New York representative.[8]

Eben and Emily spent the summer of 1906 in Ocean Park, California, as they had done in 1903 and 1904, and their children and grandchildren were frequent visitors. Smith had suffered another stroke by then, but that did not stop him from enjoying visits with his grandchildren, taking them to the local skating rink on afternoons when he felt up to it. In May, he sent Cora $5,000 so she could travel to California first class. Josephine Smith and her children arrived for a visit in late June or early July, but she had to cut her visit short when a telegram arrived at the Smith house informing her that her mother, Josephine Hill, had died. Frank, Josie, and the children left for Denver the same day. As the days of summer drifted away, Smith wrote to

John Campion that "I have not thoroughly recovered from my recent illness, but am on deck and attending to business as well as I can."[9]

Near the end of October, Eben and Emily decided to travel to Colorado to visit Cora and her children. They arrived in Colorado on October 27, and on November 3, Eben fell ill with severe stomach pains. The family summoned a number of doctors, who quickly diagnosed their patient with appendicitis. Due to Eben's advanced age and his weakened condition from the strokes, the doctors decided against operating. Two days later, on the afternoon of November 5, 1906, "as the golden November sunlight was streaming through the windows, a man whose name is known the length and breadth of the state of Colorado, entered upon that journey whose end no man knows." Eben Smith was dead from acute peritonitis at age seventy-four.[10]

News of the mining man's death spread quickly. The *Breckenridge Bulletin* called Smith one of the most "highly respected" men in mining, a sentiment echoed by the *Denver Republican*. The *Denver Post* said that to Smith was "due the perfection of the science of ore reduction," and both the *Post* and the *Denver Times* called him the "dean of mining in Colorado." All shared a sense that Colorado's mining industry had lost one of its giants. The family held Smith's funeral at his former home at 951 Logan on November 7 under the direction of Reverend Charles Marshall, rector of St. Barnabas Episcopal Church. Nona Brooks of the Divine Science Church also made remarks, and Miss Thompson sang several hymns. State officials, mining men, and numerous pioneers attended the funeral. Governor Jesse McDonald, John Campion, Robert Cary, Thomas Daly, Walter Cheesman, General George Cook, Major Lafayette E. Campbell, and David Moffat's son Fred—all longtime business and political associates—served as pallbearers. Following the funeral, Smith was buried in a temporary vault at Fairmount Cemetery while work began on a private mausoleum for the Smith family.[11]

On November 15, Frank filed Eben's will for probate in Denver County Court. At the time, estimates suggested his estate to be worth $3 million to $10 million. Half of the estate went to Emily, with the remaining half divided between Frank and Cora. In his will, Eben arranged to invest $50,000 in securities for the benefit of his sisters, Nancy Bassett and Emily Lyons, who received the interest, while Emily's sister, Sarah Rundel, received $5,000. The biggest surprise in the will was in connection with Eben's son Lemuel. Due to his "personal habits," Smith left Lemuel only $25,000, which he ordered

FIGURE 9.2. The Eben Smith mausoleum at Fairmount Cemetery. Author photo.

to be placed in trust and only the income paid to him. Eben went on to say in his will that if at any time after his death his executors determined that Lemuel had "fully and permanently reformed his life and habits and that he will not waste or squander" the money, the trust could be revoked and the money paid to him. Frank and Cora served as executors, and as a safeguard against questions about the genuineness of the will or shady attempts to substitute different pages, "upon the margin of each page of the will is written in the large, scrawling hand of the millionaire the initials 'E.S.' "[12]

Smith's will made it perfectly clear that he was first and foremost a mining man. At the time of his death, he was still part owner of nearly every major mine with which he was involved: the Maid of Erin, Henriett, Resurrection, Grey Eagle, Pocahontas, Ibex, Old Hundred, and Gazelle in Leadville and the Victor, Venture, Roxana, Little Don, and Mount Sopris in Cripple Creek. None of these mines were producing or ever would produce as they had in the past, but they would remain associated with the name of Eben Smith for many years to come. The family set up the Eben Smith Estate Company to handle the business aspects of Smith's estate, and the International Trust Company of Denver administered the corporation.[13]

At the time of his father's death, Lemuel Smith was earning $5 a day as an engineer in Leadville. He, his wife, Nellie, and their daughter, Fay, lived in "moderate circumstances"; Lemuel provided "for them comfortably" and had a "neat, attractive home." Lemuel commented only briefly on the terms of his father's will at first, saying it was not the right time to talk about it as the death of his father had been a "severe loss," but his friends hinted that he would contest the will. That was exactly what Lemuel eventually threatened to do, saying he would reveal the skeletons in the Smith family closet if it became necessary. Emily, Frank, and Cora pleaded with Lemuel not to do that, and they compromised by paying him $50,000.[14]

With the announcement of Lemuel's settlement on November 27, 1906, the public also learned that the then estimated value of Eben Smith's estate was just over $1 million, the majority of which consisted of $636,000 in annuities Smith had purchased from Mutual Life of New York in 1905. The rest of the money, according to the *Denver Times*, was "frittered away in losses during the last year of Eben Smith's life," on businesses such as the Pacific Wireless Telegraph Company and the Electrolian Organ Company. A true, final account of Smith's estate was never published, but a partial inventory completed on March 26, 1907, put the value at $1 million, with every expectation that it would grow larger. By then, Eben's sister Emily Lyons had died as well, and the family decided to divide her half of the $50,000 trust among Emily, Frank, and Cora. The last figures for the estate, published in late 1907, put it at $5 million.[15]

On November 16, just eleven days after Eben's death, a battle over the Bakersfield and Ventura Railway erupted in a Los Angeles courtroom. Shortly before his death, Eben had loaned the railroad over $200,000, which

FIGURE 9.3. Lemuel (a), Fay (b), and Nellie (c) Smith, 1906. *Denver Post*.

gave him bonds in exchange. While trying to settle his father's estate, Frank attempted to sell Eben's interest in the railroad to Senator William A. Clark, but Major H. M. Russell, president of the railroad, objected to the sale. During the first day of the court battle, US district attorney Oscar Lawler, who represented Smith's estate, even went so far as to suggest that

Russell's handing of the railroad was "responsible for the untimely death of Mr. Smith." Russell denied the charges, said that he and Smith had been on excellent terms, and accused Frank of trying to sell the railroad out from under him. The courts eventually allowed the Smith estate to foreclose on the railroad, and after an unsuccessful appeal, the line was sold to Frank on December 21, 1906, for $200,000. Estimates placed the value of the railroad at over $1 million. In 1908, a group of Colorado men led by Charles Carnahan and Eben's lawyer John Ewing incorporated a new company to manage the railroad, followed by yet another corporation in 1910. The American Beet Sugar Company bought the railroad in 1911 and continued to operate it as the Ventura County Railway.[16]

After her husband's death, Emily did not return to California. The family donated Eben's large, valuable orchid collection to the city of Los Angeles, and Emily sold the house on Seventh Avenue for $146,000 (the property was eventually subdivided, and in 1909 a developer tore down the house and built a large apartment building on the property). She bought a house in Denver at 964 Logan, across the street from her daughter, but spent most of her time at Estamere and visiting with her grandchildren, taking them on outings to Elitch Gardens and other local attractions.[17]

On November 5, 1907, exactly one year after Eben's death, all the Denver newspapers announced that Judge George Allen, the father of Josephine's former fiancé Harry Allen, had dissolved the marriage of Frank and Josephine Smith. The news shocked Denver society, and the details revealed in the divorce hearing suggested that Lemuel might not have been the only Smith son who had failed to live up to his father's high expectations. Josephine sued for divorce on grounds of cruelty, and Judge Allen agreed to hear the case before the court opened for the regular session in an effort to decrease publicity. There had been rumors of women and too much wine throughout the marriage, but Josephine made no mention of the women. After the hearing, she told reporters that Frank was as good a husband as there could be when he was not drinking, but if he drank a little, he "was not always kind to me." In court, she testified that Frank often used abusive language in speaking to her and that she spent several days in bed under a doctor's care after one especially bad episode. Frank's only comment after the hearing was that he felt bad about it but that Josephine would be "happier alone." Josephine received custody of the children, the house at 1801 York Street and

FIGURE 9.4. Emily Smith with grandson Harold Carnahan at Elitch Gardens, 1908. Michel Tritt Collection.

all its contents, a cash settlement of $50,000, and $250 a month for the care of the children.[18]

After the divorce, Frank left for California to attend to unfinished business related to his father's estate. When he returned to Colorado, he moved into his mother's house at 964 Logan. Nine months after the divorce, on August 3, 1908, Josephine married Richard Hughes, a private detective from Kansas City who had worked on her divorce case, and in the process set off a new round of trouble. Frank went back to court to get custody of his children, claiming that Josephine had acquired a drug habit. The court ruled in favor of Josephine. Frank did not remarry until December 6, 1909, when he married Lula Emma Wupperman at the Trinidad Methodist Church in Trinidad,

Colorado. Lula was a southern woman of "especial charm and beauty," whose divorce from her first husband, Texas merchant Richard Wupperman, became final on November 23, 1909—only thirteen days before she married Frank. The marriage came as a surprise to all, including Frank's sister Cora, who told a reporter from the *Denver Times* that she had known of her brother's engagement to Lula but "did not know that the marriage was to take place so soon."[19]

In December 1907, Emily Smith's doctor, William Bagot, diagnosed her with cancer and told her she had two years to live. By late 1909, her health was failing rapidly, especially after she suffered a hemorrhage in October of that year. In an effort to prolong her life, doctors operated on her on Christmas Eve 1909, but due to her age there was little hope for recovery, and Frank and Cora spent all of Christmas Day with her. That day, Emily called all her servants into her room, gave each of them checks ranging from $1,000 to $2,500, and "bade them be good and useful men and women." Her mind remained clear, but "when the angel of death came to claim her she said she was ready to go." She died on December 26 at age seventy-three. The family held private funeral services at her home on December 28, again under the direction of Reverend Charles Marshall. David Moffat served as an honorary pallbearer, while the active pallbearers were his son Fred, Robert Cary, Thomas Daly, Carl Ridgeway, William Wiggington, and General George Cook—all longtime associates of the family. After the funeral, the family buried Emily next to her husband in the completed family mausoleum.[20]

Frank continued to live in his mother's house after her death, and Cora wrote to a friend that she was upset that his new wife would get everything in the house that had belonged to her mother. The inventory of Emily's estate, completed on July 28, 1910, valued it at $201,847, although the true worth was probably higher. In addition to the mining stocks and real estate inherited from her husband, there was a great deal of jewelry included in the estate, along with the Packard Eben and Emily had bought in 1906. Frank and Cora got just over $80,000 each, and each grandchild received a $3,000 trust, payable when they reached twenty-five.[21]

After Emily's death, Frank and Lula left Denver for an extended tour of Colorado, returning on April 9, 1910. Two of Lula's four children from her previous marriage moved in with the couple, and Frank's children from his first marriage were frequent visitors to the house. On the night of May 6,

1910, Lula, her two children, her sister, and her niece gathered for dinner in the dining room and waited patiently for Frank to come downstairs. Shortly after 7:00, Marion Mehan, one of the maids, ran from the house screaming "Frank Smith has shot hisself [*sic*]." A number of neighbors, including Frank's brother-in-law Charles Carnahan, ran to the house. Several doctors responded to the family's summons, and they determined that the bullet, which entered an inch above Frank's heart, had passed completely through the body. Someone called for a car and drove Frank to St. Joseph's Hospital, where doctors reported that his condition was grave. Despite the doctors' best efforts, Frank died early in the morning of May 9, 1910, at age thirty-eight.[22]

The Smith family immediately requested an inquest because they wanted to make sure "no legal technicality" would prevent them from collecting on Frank's $300,000 life insurance policies. The family insisted that the shooting had been an accident. Family friend George W. Cook told reporters that guns had fascinated Frank ever since he was a child; that he often enjoyed taking them apart, cleaning them, and putting them back together; and that it was easy to understand how an accident could have happened. Charles Carnahan was less expansive, telling reporters that the gun had discharged in "some way unknown to anyone." All agreed that a suicide attempt was the last thing the shooting could have been, since Frank had no reason to commit suicide: he was young, he had recently married a pretty woman, and his finances were in good shape. The family finally settled on the story that Frank had laid down on his bed to clean the .38 caliber revolver he always kept beside it, and the gun became entangled in the cord of a drop light on the nightstand and fired.[23]

Denver's newspapers found the explanation less than satisfactory, in part because of the Smith family's extraordinary efforts to come up with a believable story. The coroner's jury dashed the family's hope for an accidental shooting verdict, ruling on May 10 that Frank shot himself intentionally. Particularly damaging to the case was Charles Carnahan, who testified that when he reached his side, Frank had "gasped" to him "I tried to do a good job." No one at the time gave a reason for Frank's suicide, but according to family legend he took his life after a beautiful French dressmaker who made dresses for many of Denver's high-society women rejected his advances. The family held Frank's funeral at 964 Logan on May 11 under the direction of

Reverend Charles Marshall, after which he was buried in the family mausoleum at Fairmount Cemetery.[24]

Frank's will divided his estate—valued at $265,872 without his life insurance—in half, with Lula receiving $140,000 and Frank's three sons the rest (the life insurance was also divided in half between Lula and the children). However, Claire S. Miller, an attendant at a cigar stand in a downtown Denver building, then appeared, claiming she was Frank's common-law wife. Miller told one of Frank's attorneys, Karl Steinhauer, that she had lived with Frank as his wife for many years before his divorce from Josephine and had friends in California who would recognize her as such. According to the *Los Angeles Herald*, Miller even claimed to have three children with Frank. Lula Smith denied the charge, but the *Herald* reported that the story would "bring to light some sensations" in Frank's lifestyle. Most people, including his mother, recognized that Frank was somewhat careless in his personal life. Shortly before her death, Emily Smith told her friend Frances Watson that "Frank was running the same gait as his father" and that if she could, she would "gladly lay him beside his father." But it is hard to imagine that Josephine Smith would not have mentioned Frank having a second family in her divorce suit, and on most of the occasions when Frank visited his father in Los Angeles, he had his family with him, making it unlikely that Miller would have spent enough time with him there—especially under Eben's watchful eye—to be recognized as his wife. The family wanted to be rid of the situation, though, and settled with Miller for $50,000. Fifteen months after Frank's death, Lula married Karl Steinhauer. Their marriage was short-lived, and Lula went on to marry and divorce twice more before her death in 1950 at age seventy-three.[25]

As the Smith family scandals unfolded, the Smith-Moffat mines continued to operate under the direction of David Moffat. However, the majority of Moffat's attention went to his other business ventures, including the First National Bank of Denver and the Denver Union Water Company. His most cherished project during this time, however, was the construction of his latest railroad, the Denver, Northwestern and Pacific (more commonly known as the Moffat Road); in December 1910, he traveled to New York to try to raise money for the project. The trip turned into a sad one for the financier, as his fellow traveler, Colonel Leonard Eicholts, returned to Denver where he died. Moffat soon received news of the deaths of two US senators who were close friends: Charles Hughes of Colorado (a former lawyer and mining associate of Smith and Moffat) and Stephen Elkins of West Virginia. In early March 1911, Moffat was still in New York when friends in Denver received word that he was seriously ill. By the morning of March 18, it appeared that he was recovering, as he was seen walking in the corridors of the Hotel Belmont at around 10:00. Half an hour later, he was dead at age seventy-two, although whether from illness or suicide is still debated.[26]

Under the terms of Moffat's will, his executors organized the Moffat Estate Company. On July 1, 1912, the directors of the company officially incorporated the Smith-Moffat Mines Company to take over the two men's mining interests, and the Eben Smith Estate Company became a subsidiary of the Moffat Estate Company. As the sole surviving executor of Smith's will, his daughter Cora Carnahan gained control of the estate and was issued 50,000 shares of stock in Smith-Moffat Mines while the Moffat Estate Company owned the other 50,000. Her husband, Charles Carnahan, served as one of four directors of the mining company. The Smith-Moffat Mines Company continued mining operations on most of the Smith-Moffat properties through the late 1930s. One of the problems still facing the Victor Mine in 1936 was the lost vein, still missing thirty-seven years after Eben Smith had expressed his fear that it had drifted onto the neighboring Isabella property. Realizing that Smith had been correct, in August 1936, Smith-Moffat Mines negotiated an agreement with the owners of the Isabella that would allow them to work the once-rich vein. The project turned out to be less than profitable, and by the late 1930s, it was clear that the Smith-Moffat properties were long past their prime and the company abandoned work on them. The directors of the company spent the next twenty years trying to find someone to lease the land.[27]

J. Churchill Owen, president of the Moffat Estate Company, told a business associate in the early 1970s that unlike the Moffat family, the Smith family "refused" to cooperate with the estate company. In large part, he blamed this on the fact that the Smith heirs were "scattered" all over. However, by the time Owen made that statement, there were few Smith family members left to cooperate. In the twenty years following Eben Smith's death, tragedy routinely struck the Smith family.[28]

Josephine and Richard Hughes divorced in February 1912, after Josephine complained that her husband failed to support her. Judge George Allen granted the divorce. According to Smith family legend, Josephine found herself alone and overwhelmed by having to care for three children on her own. A couple from Texas befriended her and then proceeded to steal everything they could from her before disappearing. Charles and Cora Carnahan divorced on April 3, 1915, after twenty-two years of marriage. Newspapers described Charles as one of the wealthiest mining men in the state, a dramatic change from the days when Eben Smith had to send his daughter pin money. Cora sued for divorce on grounds of desertion, saying her husband had left her about two years before, even though they had just recently announced the engagement of their daughter Doris to Courtland Dines, son of Eben's attorney Tyson Dines. It took Judge John H. Denison less than ten minutes to grant the divorce, after which Cora and her daughters left for New York. In 1916, Cora married stockbroker Thomas Costello, who was eighteen years younger than she was. Charles moved to Pinehurst, North Carolina, and in 1919 married Marie Romaine Brinkerhoff, who was thirty-three years younger than he was. They later had a daughter named Ann.[29]

Lemuel Smith and his wife, Nellie, separated in 1909, and he moved to Denver where he rented a room above a bar on Larimer Street. Nellie and their daughter Fay went to California, but the couple never divorced and Lemuel stayed in touch with his daughter for the rest of his life. Someone, possibly Lemuel himself, established a trust fund with the money he received from his father's estate, which gave him an income of about $1,000 a year. The owner of the bar above which he lived advanced him enough money every month to keep him going, and every month Lemuel arrived, neatly dressed, at the offices of the International Trust Company in Denver for his check. He always settled with the saloon keeper, then returned to his rooms, exchanged his good clothes for an old suit, and proceeded to get drunk. After struggling with alcohol for most of his adult life, Lemuel finally quit drinking in 1916 when Denver went dry. According to the *Denver Post*, he "braced up in every way," and in 1918 he bought a war bond. In mid-April 1918, he was at the International Trust Company as usual and said he would be back to buy another bond. The next day, someone found him unconscious in a room at the Windsor Hotel in Denver and took him to Denver County Hospital. He died there, without regaining consciousness, on April 21, 1918, at age fifty-nine. He was buried in the family plot at Mountain View Cemetery in Oakland, California. Nellie died in San Diego on June 7, 1934.[30]

Eben LeRoy Smith, Frank and Josephine's oldest son, was a student at the University of Colorado in 1914 when he and his fiancée, Margaretta Onderdonk, eloped in Cheyenne. When the United States entered World War I, Eben left his job at the International Trust Company in Denver and entered the officer's training corps at Camp Funston, Kansas, after which he became a first lieutenant in the army. While fighting in France in late October, Eben became so ill that he required hospitalization, but he left the hospital to rejoin his unit. On November 4, one week before the war ended, Eben was killed in action near Beaufort, France. According to his friend Lieutenant Max Melville, Eben was wounded in the head by machine gun fire and never regained consciousness. After setting aside $1,000 for the care of the Smith mausoleum at Fairmount, Eben left the bulk of his $500,000 estate to his wife. Eben's body was moved to the family tomb in 1921, his grave marked by a plaque that detailed his heroic death.[31]

In 1916, eighteen-year-old Melvin Hill Smith, Frank and Josephine's second son, was a student at Colorado State Agricultural College in Fort Collins. There, he met and married Marian Adelaide Parker and they had two children, Melvin Jr. and Betty Adelaide. Seven weeks after Betty's birth, Adelaide filed for divorce, which was granted on January 20, 1920. Estimates placed Melvin's fortune at $200,000, of which Adelaide received $20,000 plus $150 a month in child support. Four days after the divorce, Melvin married his second wife, Virginia Stone, and the two left for Los Angeles. In California, Melvin liked to spend time around police stations, often driving officers in his Locomobile when they were on a case. Just before daybreak on the morning of June 28, 1921, Melvin, Los Angeles police officer Jack H. Abbott, and Abbott's friend Jean Bruneault were riding in Melvin's Locomobile. Melvin attempted to swing the car around a corner at a high rate of speed in a residential section of Los Angeles, but the car skidded, struck a stone pillar, and was completely demolished. Abbott and Melvin, who was twenty-three, were dead by the time people reached the scene, but Jean was only "slightly bruised." Melvin was also buried in the family mausoleum. His and Virginia's son, William Eben Smith, was born on July 4, six days after his father's death.[32]

Frank Leonard Smith (who went by Leonard), Frank and Josephine's youngest son, and his wife, Rosina, lived in Santa Fe and had two children, Frank Jr. and Rosina Josephine. Leonard and Rosina divorced in February 1924, and two months later Leonard was in Colorado visiting his aunt Cora and cousin

Harold Carnahan at Estamere when he fell ill with sleeping sickness. Doctors took him to Mercy Hospital in Denver, and according to the *Denver Post*, "every means known to medical science . . . was resorted to in an effort to save the young man's life," but it was not enough. After being in a coma for twelve days, Leonard died on April 26, 1924, at age twenty-four. A year and a half after his death, the family had Leonard's body moved to the family mausoleum. The families of the three boys later unsuccessfully sued Cora Costello to get the trust funds Emily Smith had established for them in her will in 1909.[33]

Charles and Cora Carnahan's son Harold was attending Yale University at the outbreak of World War I, at which time he quit school and enlisted in the army. After the war, Harold returned to Colorado, but he suffered from melancholia and frequently sought treatment at sanitariums. In the spring of 1924, Harold spent several weeks with his mother and cousin Leonard at Estamere in the hopes that this would help lift his spirits. To those who knew him it appeared to work, as he showed signs of improvement, but when he returned to Denver, his condition worsened. In early November, he left for Pasadena, California, where he checked into the Las Encinas Sanitarium. There, on the night of November 9, Harold, who was thirty, shot himself. His sister Emily's husband, Charles Wilson, went to California to accompany Harold's body back to Denver for burial in the family mausoleum.[34]

With the deaths of Eben, Melvin, and Leonard Smith and Harold Carnahan, the only surviving heirs of Eben Smith were his daughter Cora and her daughters Doris and Emily and Lemuel's daughter Fay. Although Charles and Cora Carnahan had divorced in 1915 and Charles and his second wife, Marie, moved to South Carolina, he remained a director of Smith-Moffat Mines until 1929. Four year later, however, the directors of the company called on him, as one of the surviving directors of the Venture Company, to sell that property to the mining company. It was his last job for Eben Smith. In October 1935, the Carnahans were visiting Asheville, North Carolina, when Marie fell ill. She died in Asheville on October 8, 1935, at age forty-one and was buried in Riverside Cemetery in Asheville. Charles died in Asheville two years later, on August 20, 1937, at age seventy-six and was buried next to Marie.[35]

Josephine Smith, perhaps the most tragic figure in the Smith family, had moved to Santa Fe with her son Leonard in 1920. She bought a house on Palace Avenue next door to the sister of Rosina Bergere, Leonard's future wife. After Leonard's death, she traveled a great deal and rented a house in Denver as needed. She spent her last years in a nursing home in Denver, where her granddaughter Betty Works remembered visiting her from time to time. Josephine died in Denver on January 5, 1946, at age seventy-three. After her funeral at St. John's Evangelist Catholic Church, she was buried in the Smith mausoleum.[36]

Eben Smith's last surviving child, his daughter Cora, stayed in Denver for the most part, but she did spend some time living in New York City and Arizona as well as traveling. She and Thomas Costello separated in late 1919 or early 1920 but had reunited by 1924. In 1928, she listed herself as his widow in the Denver City Directory, although he was alive and living in Los Angeles (he died there in 1968 at age eighty). Cora's daughter Doris and Courtland Dines were also divorced by 1920, and Doris later married Robert Van Schaack, whose brother Henry founded a well-known Denver real estate company. In 1926, Cora moved out of her father's old house at 951 Logan and into a mansion she built at First and Race, across the street from the Denver Country Club. Her one business venture was in the mid-1920s when she sold Chinese art goods at the Brown Palace Hotel. Betty Works remembered being picked up by Cora's chauffeur, John, to have tea with her when Betty was about ten years old. Cora also gave Betty and her brother Melvin Jr. books to read, including a first edition of *The Wizard of Oz*.[37]

To keep the memory of her mother and father alive, Cora named two chairs in their honor at the Central City Opera House, but for the most part she lived quietly. To some of her longtime friends, such as Fanny Watson O'Kane, she commented that she was unhappy, which was not surprising considering that between 1906 and 1928 she had experienced the deaths of her parents, the suicides of her brother and her son, the untimely deaths of her three nephews, and the failure of her two marriages. In 1951, the list of tragedies increased when one of her grandsons, Peter Van Schaack, committed suicide. In 1954, Cora moved into the Salve Regina Convalescent Home at 975 Pennsylvania, where she died on January 17, 1956, at age eighty-six. After a private funeral, Cora was buried in the family mausoleum. The next year, on November 14, 1957, her daughter Doris died at age sixty-two.[38]

After Cora's death, her shares in Smith-Moffat Mines passed to her daughters Doris and Emily Wilson, and after her death, Doris's shares went to her estate. Smith-Moffat Mines had sold the Resurrection in 1942 for $10 but still owned all the other mines. By the mid-1950s, Smith-Moffat Mines was relying on the stockholders to pay the taxes on the properties, and the company's directors were desperate to get rid of the mines. In 1959, J. Churchill Owen hired engineers to survey the remaining Smith-Moffat properties and offer suggestions as to what the company should do with them. The engineers, in their final report, recommended getting rid of all the mines. On October 7, 1959, Smith-Moffat Mines sold all the Cripple Creek property, including the Victor for which Smith and Moffat paid $65,000 in 1892, to the Golden Cycle Mining Company for $10. At the time, the company owed the Smith estate $8,600, Emily Wilson $6,200, and the estate of Doris Van Schaack $5,500. Most of the Leadville properties were disposed of around this time, although who bought them and what they paid for them is unknown. The last mine sold was the Old Hundred property, which Smith had bought in 1903; the company sold it to the Climax Molybdenum Mine in 1961. Smith-Moffat Mines shut down in 1962, and the few remaining assets became property of the Moffat Estate Company.[39]

Eben Smith's last surviving grandchildren were Cora's daughter Emily and Lemuel's daughter Fay. Emily Carnahan Wilson, who preserved her grandfather's papers and donated them to the Denver Public Library shortly before her death, was a colorful character who was best known in the late 1940s and 1950s as the owner of the famous Glory Hole Tavern in Central City. Emmy, as she was known, died in Denver on August 4, 1963, at age sixty-one. Fay Smith Landweer and her husband, Herman, lived in California, where she worked for San Diego State University for many years before retiring in 1965. Fay died in 1976, five years after the Moffat Estate Company announced that it was liquidating all assets, including any remaining assets from Smith-Moffat Mines. The company shut down in 1972, 112 years after Eben Smith first arrived in Colorado.[40]

According to historians Carl Abbott, Stephen J. Leonard, and David McComb, the gambler was a dominant figure in the early reports of life in Colorado mining camps. Monte and faro dealers entertained customers in saloons in Denver and Golden in the winter of 1859, with six tables in the Denver House alone kept busy twenty-four hours a day. But card players, they write, were not the only gamblers in the territory. Everyone who came west seeking their fortune, "miner and merchant, barkeeper and hardware salesperson, editor, land dealer, and politicians . . . each risked his or her future on the wager that the richness of Rocky Mountain ores and the openness of the new society would compress into a few years the career that would have required a lifetime of hard work in the established East." Far from a ruthless capitalist like Rockefeller or Carnegie, Eben Smith gambled his future on the mines of the West, and like the westerners of the Gilded Age who Gene Gressley writes about, he was constantly searching for a brighter future for himself and his family. Gressley argues that most westerners' "essential economic desires were basically for more blessings than [they] possessed today." Growing up with a large family on a farm in Pennsylvania did not seem to offer him the brighter future Smith sought for himself, so he bet on finding that future in the California Gold Rush. He then gambled a fortune made in California to bring a stamp mill to Colorado, and he was always certain that the next mine he purchased or became involved with would be the next big producer and that a greater fortune was there to be found. When his business ventures near the end of his life constantly drained his financial resources, he remained convinced that a little more money, harder work, and the right people would turn the situation around.[41]

When the situation called for it, Smith could sound like the typical booster trying to oversell a mine to eastern or European investors, but he seemed to prefer keeping control over his mines rather than letting outsiders in. However, he also knew that developing and working the mines required a great deal of money, and investors were a good source of that money. Smith loved to dabble in the stock market himself, so he did what was expected of him. The end goal, though, was to boost his mining career. He could be ruthless when he believed the situation called for it, as seen in his response to striking miners in 1894 and 1896, but he could also be far too trusting, which led some of his closest associates and employees to take advantage of him. Through it all, though, his primary motivation was to secure more blessings for tomorrow. He was an optimistic gambler, constantly winning and losing but hoping to come out ahead in the end.

No industry played a more important role in the development of Colorado in the nineteenth century than mining. Eben Smith, with his knowledge of and expertise in the field, played a key role in the growth of mining and, as a result, in the growth of Colorado. He was a fast learner, and his experience in California placed him in the "front rank as a mining engineer" by the time he arrived in Colorado. Smith's ability to determine that a stamp mill would work in Colorado based solely on reports he heard from those who had been there says a great deal about his abilities as both a miner and a milling expert. As the *Denver Republican* said in 1886, the Smith and Chaffee Stamp Mill deserved credit for developing the mines of Gilpin County and prodding the mining industry of Colorado to new heights. While academically trained engineers would come dominate the field one day, the earliest miners relied on men like Smith who had learned from experience.[42]

While Smith's accomplishments in Central City were impressive enough, he impacted mining throughout the state during his career. In Leadville, his Grey Eagle and Pocahontas Mining Company was the first deep-mining operation in the district. Facing the common problem of rising water in the mine shafts, Smith responded by putting in pumps and draining the water so work could continue. A number of deep-mining operations in Leadville, facing the same water problem, followed Smith's lead after witnessing his success. Several of Smith's obituaries referred to him as a pioneer of the Cripple Creek area and described Smith and Moffat's Victor Mine as the "first real money-maker of the district." Throughout his life, he had the luck and ability to be on hand when a major and profitable development in mining took place, and he possessed the knowledge to take full advantage of it. Of course, Smith was also fortunate in his business partners. Jerome Chaffee and David Moffat had wide circles of influence in politics, banking, and the law—all of which served their mining interests well. Both men trusted Smith and his management of their mines, and the fact that he was often also an owner only increased his own desire to have the mines do well.[43]

Smith's contributions to Colorado's mining industry went beyond the mines themselves, sometimes in ways he did not like. He and many of his fellow mine owners adopted labor policies that allowed unions to gain a foothold in the state. The Smith-Moffat mines were not unusual in the wages they paid and the hours they expected miners to work. In fact, most mine owners considered Winfield Scott Stratton, who quickly settled with the striking miners in 1894, unusual. Smith was clearly upset by the strikers interfering with his business, but much of what he said was also a case of his bark being worse than his bite. It is highly unlikely that he would have given the final order to blow up a bridge on the Florence and Cripple Creek Railroad because it would have been expensive to replace the bridge and the road would have lost money while it was shut down for repairs. Smith also made sure that the 150 guns he helped buy during the strike were always locked up, guaranteeing that no one would be injured or killed by them. Smith had the typical nineteenth-century businessman's hostile views toward organized labor, and both sides were fighting for what they believed to be their rights, whether as a mine owner and manager trying to operate his mines or as a mine laborer trying to earn a better living.

Eben Smith's obituary in the *Denver Post* stated, "So long as mining may endure his name, his brain, and his work will be an integral part of mining." During his lifetime, he earned millions of dollars being involved in a business he loved. While some of the things he did, such as his participation in David Moffat's scheme to rid the Victor Mine of its debts in 1900, would now be considered unethical if not outright illegal, he did not do anything his fellow mine owners would not have approved of or taken part in themselves at the time. Smith and his family enjoyed the benefits his wealth provided, and for the most part, he was generous to others with his wealth. Perhaps as a testament to his financial abilities, Smith also managed to hold on to much of his fortune, unlike many of his associates.[44]

Mining in Colorado has indeed endured, but the state's relationship with the industry Smith helped create has been complex, particularly in the twentieth century as the industry lost its dominance in the state. Beginning in the 1960s and 1970s, the environmental impact mining had on the state became a great concern. National legislation impacted both former and current mining sites in Colorado, and residents around the few active operations were increasingly aware of the effect those operations had. Even cleanup efforts by the Environmental Protection Agency could have negative impacts, as in 2015 when 3 million gallons of wastewater in the Gold King Mine near Silverton were accidentally released into the Animas River, once again prompting people to debate the state's mining history and the environmental impact it continues to have.[45]

At the same time people began discussing the environmental impact, the state's former mining sites became big business through heritage tourism. Tourists could visit mines in Central City, Idaho Springs, Cripple Creek, and Silverton, among other places. One of those mines was Smith's Old Hundred Mine near Silverton. Visitors to Silverton could also ride the Durango and Silverton Railroad, walk through a fully restored mill, and admire the architecture of the town. Central City, Black Hawk, and Cripple Creek turned to casinos in the 1990s to help revitalize those towns, with money from the industry going toward historic preservation throughout the state. The largest gold mining operation in Colorado, the Cripple Creek and Victor Gold Mining Company, helped fund the restoration of the town of Victor as part of its operation.[46]

Smith's involvement in railroads and other transportation projects, politics, and banking were important to the development of the state as well. The First National Bank of Denver, for example, was in existence for more than 125 years and played "a central role in the growth and development of Denver, the State of Colorado, and the Rocky Mountain Region." Of course, not everything Smith touched was successful, and his confidence in his abilities led to a few failures. A few mining properties deceived him, such as the Tam O'Shanter in Ashcroft, but more often his less than spectacular business ventures were outside mining. His willingness to gamble in hopes of increasing his fortune caused him to lose a great deal of money with the Los Angeles Art Organ Company and the Pacific Wireless Telegraph Company. His determination to make these businesses successful, in part because of his love for business and in part to increase his own wealth, only led to frustration and financial drain on his part. Had he lived longer, he may have made a success of the wireless business but probably not the organ company, which declared bankruptcy shortly after his death.[47]

When the neighbors of the Denver Mine and Smelter Supply Company warehouse invoked Eben Smith's name in 2012 in their effort to save the building, they showed just how connected he still is to the state's mining history. Mining and its associated industries shaped Colorado and its development, and Smith had a major role in that industry. When development threatened the Mine and Smelter Supply warehouse, Denver was beginning an extraordinary period of change that would see tremendous growth and development, and it was somewhat ironic that those who lived around the warehouse hoped to use the name of someone who had spent much of his career developing Denver and Colorado to try to stop it. Despite his many business ventures, Eben Smith's first love was always mining. He rarely called himself a mine owner or a general manager, instead referring to himself simply as a "miner." He preferred to work behind the scenes while David Moffat or Jerome Chaffee filled the role of promoter. It was his knowledge and his experience, though, that contributed a great deal to making mining a major industry in Colorado, leading to all the benefits of economic growth and development. In 1906, the *Denver Times* wrote that Eben Smith was "a pathfinder in the wilderness that existed before the statehood days and one of the builders of this great commonwealth," while the *Denver Post* said he "developed Colorado." All agreed, however, that his greatest contributions were in mining, which was what he would have hoped for.[48]

FIGURE 9.5. Eben Smith with his granddaughter Emily Carnahan Wilson, about 1904. Author's collection.

Notes

Abbreviations

BV: Bound volume
DP: *Denver Post*
DR: *Denver Republican*
DT: *Denver Times*
FF: File folder
RMN: *Rocky Mountain News*
RMS: *Rocky Mountain Sun*
WHGD: Western History and Genealogy Department, Denver Public Library

Introduction

1. Jeremy P. Meyer, "Denver Neighbors Crusade for Historic LoDo Warehouse," *DP* (July 5, 2012): 1A, 8A; Vincent Carroll, "An Abuse of Historic Status in Denver," *DP* (July 11, 2012): 25A.

2. "Eben Smith," Eben and Emily Smith Files, Gilpin History Museum, Central City, CO.

3. Steven F. Mehls, *David H. Moffat, Jr.: Early Colorado Business Leader* (New York: Garland, 1989), 48; Norman E. Tutorow, *Leland Stanford: Man of Many Careers* (Menlo Park, CA: Pacific Coast Publishers, 1971), 24; Duane A. Smith, *Horace Tabor: His Life and the Legend* (Boulder: Colorado Associated University Press, 1973), 71–72; John J. Lipsey, *The Lives of James John Hagerman* (Denver: Golden Bell, 1968), 21–23, 155; John H. Davis, *The Guggenheims: An American Epic* (New York: William Morrow, 1978), chapters 4–6.

4. Mehls, *David H. Moffat, Jr.*, 155–156; Tutorow, *Leland Stanford*, 24; Lipsey, *Lives of James John Hagerman*, Part Two.

5. John Moring, *Arthur Hill: Western Actor, Miner and Law Officer* (Manhattan, KS: Sunflower University Press, 1994), 2, 18, 31–35; Frank Waters, *Midas of the Rockies: The Story of Stratton and Cripple Creek* (New York: Covici, Friede, 1937), 121–122, 124–125; information on Thomas F. Walsh from John Stewart, *Thomas F. Walsh: Progressive Businessman and Colorado Mining Tycoon* (Boulder: University Press of Colorado, 2007).

6. Information on Henry Teller from Duane A. Smith, *Henry M. Teller: Colorado's Grand Old Man* (Boulder: University Press of Colorado, 2002).

7. Information on Peter McFarlane from H. William Axford, *Gilpin County Gold: Peter McFarlane 1848–1929, Mining Entrepreneur in Central City, Colorado* (Chicago: Sage Books, 1976).

8. H. W. Brands, *The Age of Gold: The California Gold Rush and the New American Dream* (New York: Doubleday, 2002), 6–7.

9. Gene M. Gressley, *West by East: The American West in the Gilded Age* (Provo, UT: Brigham Young University Press, 1972), 6, 2–3.

10. Duane A. Smith, *The Trail of Gold and Silver: Mining in Colorado, 1859–2009* (Boulder: University Press of Colorado, 2009), 118; Richard White, *"It's Your Misfortune and None of My Own": A New History of the American West* (Norman: University of Oklahoma Press, 1993), 89, 191.

Chapter 1: "I Left Pennsylvania at 19"

1. Smith to Clinton, August 15, 1902, BV 10, Eben Smith Papers, WHGD.

2. Sarah Wilson Smith, *A Brief History and Complete Record of the American Descendants of John and Mary Smith* (Privately Published, 1892), vi–viii, 36, Michel Tritt Private Collection.

3. Listing for William Smith, Wayne Township, Erie County, PA, US Bureau of the Census, *Seventh Census of the United States, 1850* (Washington, DC: Government Printing Office); "Death of Eben Smith," *Erie* (Pennsylvania) *Daily Times*

(November 9, 1906): 14; William Newton Byers, *Encyclopedia of Biography of Colorado* (Chicago: Century Publishing and Engraving Company, 1901), 239–241; "Major Smith, of the Second Colorado," *RMN* (November 16, 1864): 2; Fletcher Jordan, "Jerome Chaffee and Eben Smith," Eben Smith Manuscript Series [hereafter Smith MSS], Stephen H. Hart Library, History Colorado, Denver; "Eben Smith," Eben and Emily Smith Files; Smith, *A Brief History and Complete Record of the American Descendants of John and Mary Smith*, 36.

4. Brands, *Age of Gold*, 43–44, 123, 128–129.

5. *History of Clear Creek and Boulder Valleys, Colorado* (Chicago: O. L Baskin, 1880), 680–681; Brands, *Age of Gold*, 72–74.

6. Brands, *Age of Gold*, 78, 80–82, 91; "The California Gold Rush," https://goldrushofcalifornia.weebly.com/travel-routes.html, last accessed May 10, 2019.

7. Jordan, "Jerome Chaffee and Eben Smith," Smith MSS; "From the Interior," *Sacramento Daily Union* (May 10, 1852): 3; "Another Disastrous Fire—French Corral in Ashes," *Sacramento Daily Union* (July 11, 1853): 2.

8. Ralph Mann, *After the Gold Rush: Society in Grass Valley and Nevada City, California, 1849–1870* (Stanford, CA: Stanford University Press, 1982), 27–28, 72.

9. James E. Fell Jr., *Ores to Metals: The Rocky Mountain Smelting Industry* (Boulder: University Press of Colorado, 2009), 6–7.

10. Brands, *Age of Gold*, 220, 194–195, 234, 466.

11. "Jerome Chaffee and Eben Smith," Smith MSS; Listing for George N.[W.] Jordan, death in 1850, *Iowa Mortality Schedules, 1850–1880*, familysearch.org, last accessed April 17, 2019; Listing for Caroline Jordan, Dubuque County, Iowa, *Seventh Census of the United States, 1850*.

12. Gressley, *West by East*, 28–29; Mann, *After the Gold Rush*, 17–23; Brands, *Age of Gold*, 245, 249–250.

13. Mann, *After the Gold Rush*, 40–41.

14. "By the Alta Telegraph Line," *Sacramento Daily Union* (May 24, 1854): 2; Jordan, "Jerome Chaffee and Eben Smith," Smith MSS; Mary McRoberts, *Boulder County, Colorado Censuses*, January 1992, listings for Eben Smith and Lemuel Smith, Carnegie Branch, Boulder County Public Library, Boulder, CO; Byers, *Encyclopedia of Biography of Colorado*, 239–241; *History of Clear Creek and Boulder Valleys*, 680–681; "Major Smith, of the Second Colorado."

15. Jordan, "Jerome Chaffee and Eben Smith," Smith MSS; Walter Williams, ed., *A History of Northwest Missouri*, vol. 1 (New York: Lewis, 1915), 368; "Jerome B. Chaffee's Death," *DR* (March 10, 1886): 1.

16. "Jerome B. Chaffee's Death," 1; W. A. Whitney and R. I. Bonner, *History and Biographical Record of Lenawee County, Michigan, Volume I, 1879* (Adrian, MI: W. Stearns, 1879–1880), 120–123.

17. Jordan, "Jerome Chaffee and Eben Smith," Smith MSS; Byers, *Encyclopedia of Biography of Colorado*, 239–241; Carl Abbott, Stephen J. Leonard, and David McComb, *Colorado: A History of the Centennial State*, 3rd ed. (Niwot: University Press of Colorado, 1994), 51.

18. Jordan, "Jerome Chaffee and Eben Smith," Smith MSS; Byers, *Encyclopedia of Biography of Colorado*, 239–241.

19. Jordan, "Jerome Chaffee and Eben Smith," Smith MSS; Byers, *Encyclopedia of Biography of Colorado*, 239–241; *History of Clear Creek and Boulder Valleys*, 680–681; "Major Smith, of the Second Colorado."

Chapter 2: "One of Colorado's Headest Men"

1. Abbott, Leonard, and McComb, *Colorado*, 50–52.

2. Abbott, Leonard, and McComb, *Colorado*, 57; *Rocky Mountain News* cited in Alan Granruth, *Mining Gold to Mining Wallets: Central City, Colorado 1859–1999* (Central City: Gilpin Historical Society, 1999), 3–4.

3. Abbot, Leonard, and McComb, *Colorado*, 59; Granruth, *Mining Gold to Mining Wallets*, 9–11.

4. Granruth, *Mining Gold to Mining Wallets*, 6–9.

5. Granruth, *Mining Gold to Mining Wallets*, 19–20; Listing for Ebin E. Smith, Elwood Washington Township, Doniphan County, Kansas Territory, US Bureau of the Census, *Eighth Census of the United States, 1860* (Washington, DC: Government Printing Office).

6. Byers, *Encyclopedia of Biography of Colorado*, 239–241 (first quotation); *History of Clear Creek and Boulder Valleys*, 680–681; Jordan, "Jerome Chaffee and Eben Smith," Smith MSS; "Daily News," *RMN* (March 5, 1862): 3; "Jerome B. Chaffee's Death," 1.

7. "Daily News," *RMN* (March 5, 1862): 3; "Daily News," *RMN* (May 14, 1862): 3; Jordan, "Jerome Chaffee and Eben Smith," Smith MSS; *History of Clear Creek and Boulder Valleys*, 680–681; White, *"It's Your Misfortune and None of My Own,"* 258; Smith to George Mabee, September 24, 1900, BV 6, Smith Papers.

8. *History of Clear Creek and Boulder Valleys*, 253–254; Byers, *Encyclopedia of Biography of Colorado*, 239–241; Eugene H. Adams, Lyle W. Dorsett, and Robert S. Pulcipher, *The Pioneer Western Bank: First of Denver, 1860–1980* (Denver: First Interstate Bank of Denver, 1984), 24–25.

9. Josiah M. Ward, "The Empire Builders of Colorado: David H. Moffat, the Banner Bearer," *DP Magazine Section* (October 3, 1920): 2; "David H. Moffat Is Dead: End Comes Suddenly in a New York Hotel," *RMN* (March 19, 1911): 1, 6.

10. "David H. Moffat Is Dead," 1, 6; Ward, "Empire Builders of Colorado," 2.

11. Thomas Maitland Marshall, ed., *Early Records of Gilpin County, Colorado, 1859–1861* (Denver: W. F. Robinson Printing Company, 1920), 272, 274–275; Duane A. Smith, *The Birth of Colorado: A Civil War Perspective* (Norman: University of Oklahoma Press, 1989), 27.

12. "Daily News," *RMN* (December 12, 1865): 1; "Daily News," *RMN* (October 17, 1865): 1; "Union State Convention," *RMN* (October 25, 1865): 2; "Call for a Territorial Convention," *RMN* (July 22, 1865): 1; Abbott, Leonard, and McComb, *Colorado*, 76–77.

13. "Proceedings of Union Convention," *Daily Miner's Register* (October 19, 1865): 2; "Union State Convention," *RMN* (October 25, 1865): 2.

14. *Council Journal of the Legislative Assembly of the Territory of Colorado—Fifth Session* (Central City: David C. Collier, Printer, Miners' Register Office, 1866), 7, 11, 14, 39.

15. *Council Journal of the Legislative Assembly of the Territory of Colorado—Fifth Session*, 24–25, 55, 64–65, 102, 123, 157, 158; *Council Journal of the Legislative Assembly of the Territory of Colorado—Sixth Session* (Central City: David C. Collier, Printer, Miners' Register Office, 1867), 24–25.

16. Incorporation Papers, January 11, 1867, Oversize FF 1, Georgetown and Breckenridge Wagon Road Company Papers, WHGD; Christine Bradley and Duane Smith, *The Once and Future Silver Queen of the Rockies: Georgetown, Colorado and the Fight for Survival into the Twentieth Century* (Louisville: University Press of Colorado, 2019), 61.

17. Jordan, "Jerome Chaffee and Eben Smith," Smith MSS (second quotation); "Obituary for Nellie Jeffery," *Sacramento Bee* (November 11, 1896), photocopy provided by Sacramento Public Library; Smith Family Plot, Plot 25 Lot 5, Mountain View Cemetery, Oakland, CA; "Daily News," *RMN* (February 3, 1866): 1; Smith, *Brief History and Complete Record of the American Descendants of John and Mary Smith*, 40–43; "Major Smith, of the Second Colorado"; "Dinner to Ex-Governor Gilpin," *Daily Commonwealth* (Denver) (March 25, 1864): 2. The reason Caroline and Eben divorced is unknown, as is the exact date, as no legal documents appear to exist. For more information on James Nelson Smith and the Battle of the Little Blue, see Richard Josiah Hinton, *Rebel Invasion of Missouri and Kansas* (Chicago and Leavenworth, KS: Church and Goodman and T. W. Marshall, 1865), 330.

18. "Daily News," *RMN* (January 13, 1866): 4; "Daily News," *RMN* (July 13, 1866): 4.

19. Axford, *Gilpin County Gold*, 49–51; Fell, *Ores to Metals*, 10–17, 22–24; Granruth, *Mining Gold to Mining Wallets*, 26.

20. "The Narraganset Works," *RMN* (February 5, 1866): 1; "Mining Items," *RMN* (March 1, 1866): 4.

21. "Daily News," *RMN* (February 3, 1866): 1; "Widow of Pioneer Mining Man Dead after Brief Illness," *DT* (December 27, 1909): 16; "Pioneer Woman Will Be

Laid to Rest Today: Simple Services to Mark Passing of Mrs. Eben Smith," *DR* (December 28, 1909): 12; "Death Claims Aged Widow of Late Eben Smith," *DP* (December 27, 1909): 3; Eben Smith mausoleum, Block 3, Fairmount Cemetery, Denver, CO; Tom Noel, ed., *Glory That Was Gold* (Central City: Central City Opera House Association, 1992), 266; "Daily News," *RMN* (March 6, 1866): 4; "Daily News," *RMN* (August 6, 1866 [second quotation]): 4; "Welcome to the Eben Smith House at 108 Casey," Eben and Emily Smith Files; "Obituary for Kate May Smith," *RMN* (August 27, 1868): 4.

22. "Georgetown," *RMN* (May 14, 1867): 1 (first quotation); "Daily News," *RMN* (June 11, 1867): 1; Jordan, "Jerome Chaffee and Eben Smith," Smith MSS; "Daily News," *RMN* (November 21, 1867): 3.

23. "Our Endorsements," *Daily Miner's Register* (August 11, 1867): 1; "Election Results," *Daily Miner's Register* (August 15, 1867): 1; "Daily News," *RMN* (July 24, 1867): 4; "Daily News," *RMN* (October 15, 1867): 1; "Premium List of the Agricultural Society of the Colorado Territory," *RMN* (August 30, 1867): 2; "Daily News," *RMN* (January 2, 1868): 1.

24. "Obituary for Kate May Smith," 4; "Obituary for Kate May Smith," *Central City Register* (August 25, 1868): 1; undated and untitled clipping in Eben and Emily Smith Files; Eben Smith mausoleum.

25. "Jerome B. Chaffee's Death," 1; Byers, *Encyclopedia of Biography of Colorado*, 239–241.

26. "Silver Mines of Lincoln and Bross Mountains," *RMN* (March 16, 1873): 4; Abbott, Leonard, and McComb, *Colorado*, 386.

27. Smith, *Trail of Gold and Silver*, 123–125.

28. "Record of Appointment of Postmasters," 1832–September 30, 1971, Roll 14, Colorado, Colorado Territory, Adams–Kit Carson Counties, National Archives Microfilm Publications, Washington, DC; "Jerome B. Chaffee's Death": 1.

29. Duane A. Smith, *Silver Saga: The Story of Caribou, Colorado* (Boulder: Pruett, 1974), 2, 4–8; "The Poor Man," *DR* (March 2, 1882): 6.

30. Smith, *Silver Saga*, 10, 23, 25, 28–29, 32–33; "Caribou: A Sadly Neglected Mine and Another Which Pays," *RMN* (July 19, 1876): 4; "Mining Intelligence," *DR* (May 4, 1881): 2.

31. Smith, *Silver Saga*, 63, 66; "Mining Intelligence," 2.

32. Smith to Moffat, December 9, 1876, First National Bank of Denver Papers, Stephen H. Hart Library, History Colorado, Denver.

33. Smith, *Silver Saga*, 69–70.

34. Smith, *Silver Saga*, 70–74; "Mining Intelligence," 2; December 15, 1879, clipping on visit, Eben and Emily Smith Files.

35. "Thrown into a Snow Bank: A Perilous Journey by the Superintendent of the Caribou Mine," *RMN* (December 17, 1879): 8.

36. "Thrown into a Snow Bank"; Smith, *Silver Saga*, 75–79; *History of Clear Creek and Boulder Valleys*, 680–681.

37. "Mining Intelligence," 2.

38. Sanford Charles Gladden, *Improvements in Boulder County*, prepublication photocopy (Boulder: Carnegie Branch, Boulder Public Library, 1987), 634.

39. "Daily News," *RMN* (June 16, 1881): 2; "Broken Is the Golden Bowl, the Spirit Flown Forever," *RMN* (November 6, 1906): 1.

40. McRoberts, *Boulder County, Colorado Censuses*, listings for Eben Smith and Lemuel Smith; *Boulder County Herald* quoted in Gladden, *Improvements in Boulder County*, 273, 289; "Boulder," *RMN* (May 7, 1881): 5.

41. Moffat to Smith, December 20, 1902, Box 1 FF 48, Smith Papers; "Eben Smith Dead," unknown source, Eben Smith Clipping File, WHGD; Brands, *Age of Gold*, 443.

Chapter 3: "David Moffat Set You on Your Feet"

1. Smith, *Horace Tabor*, 190–191.

2. Smith, *Horace Tabor*, 191–192.

3. "Ashcroft," *RMS* (August 19, 1882): 2; "Ashcroft," *RMS* (August 26, 1882): 2; "Ashcroft," *Leadville Daily Herald* (October 22, 1882): 3.

4. "Ashcroft," *RMS* (September 2, 1882): 2; "Ashcroft," *RMS* (October 21, 1882): 2; "Ashcroft" (October 22, 1882): 3; "Ashcroft," *RMS* (December 2, 1882): 2.

5. Byers, *Encyclopedia of Biography of Colorado*, 239–241; Smith, *Horace Tabor*, 274.

6. Don L. Griswold and Jean Harvey Griswold, *History of Leadville and Lake County Colorado: From Mountain Solitude to Metropolis*, vols. 1 and 2 (Boulder: Colorado Historical Society and University Press of Colorado, 1996), 362, 1343; Byers, *Encyclopedia of Biography of Colorado*, 239–241; "Death of Eben Smith," *Breckenridge Bulletin* (November 10, 1906), Eben Smith Clipping File; Osborn to Smith, January 18, 1906, Box 5 FF 22, Smith Papers; "Eben Smith Dead," unknown source, Eben Smith Clipping File.

7. Mehls, *David H. Moffat, Jr.*, 152–153; Gressley, *West by East*, 30.

8. Carl Abbott, Stephen J. Leonard, and Thomas J. Noel, *Colorado: A History of the Centennial State*, 5th ed. (Boulder: University Press of Colorado, 2013), 103–104.

9. Griswold and Griswold, *History of Leadville and Lake County Colorado*, vol. 1, 113; vol. 2, 1724–1725 (quotation on 1725).

10. "Jerome B. Chaffee's Death,": 1; Griswold and Griswold, *History of Leadville and Lake County Colorado*, vol. 2, 1806; Mehls, *David H. Moffat, Jr.*, 104.

11. *Carbonate Chronicle* (Leadville) (January 24, 1887), quoted in Griswold and Griswold, *History of Leadville and Lake County Colorado*, vol. 2, 1832; "To Work the Henriett," *RMN* (June 12, 1887): 1; *Evening Chronicle* (Leadville) (September 28, 1888), quoted in Griswold and Griswold, *History of Leadville and Lake County Colorado*, vol. 2, 1978.

12. *Leadville Herald Democrat* (October 12, 1889), quoted in Griswold and Griswold, *History of Leadville and Lake County Colorado*, vol. 2, 2000 (first and second quotations); Smith to McNeil, October 19, 1895, Box 5 FF 11, Smith Papers (third quotation).

13. Griswold and Griswold, *History of Leadville and Lake County Colorado*, vol. 2, 1990–1991; *Leadville Herald Democrat* (October 6, 1889), quoted in Griswold and Griswold, *History of Leadville and Lake County Colorado*, vol. 2, 2069–2070.

14. Griswold and Griswold, *History of Leadville and Lake County Colorado*, vol. 2, 1940, 1951; *Leadville Herald Democrat* (January 5, 1890), quoted in Griswold and Griswold, *History of Leadville and Lake County Colorado*, vol. 2, 2125; Fanny Watson to Teddy, March 11, 1893, Watson and O'Kane Family Letters; "Death of Mr. Smith," *Carbonate Chronicle* (Leadville) (June 20, 1887): 7; Listing for Samuel Smith, California, Oakland, Mountain View Cemetery Records, 1857–1973, familysearch.org, last accessed February 19, 2020; Smith Family Plot, Mountain View Cemetery.

15. Griswold and Griswold, *History of Leadville and Lake County Colorado*, vol. 2, 1637–1638, 1820; *Mines and Mining Men of Colorado: Historical, Descriptive and Pictorial* (Denver: John G. Canfield, 1893), 17.

16. *Leadville Herald Democrat* (January 1, 1892), quoted in Griswold and Griswold, *History of Leadville and Lake County Colorado*, vol. 2, 2008–2009 (second quotation); *Leadville Herald Democrat* (November 28, 1893), quoted in Griswold and Griswold, *History of Leadville and Lake County Colorado*, vol. 2, 2022 (first quotation); Check from Smith to Mrs. J. J. Brown, August 28, 1898, Box 17, Smith Papers.

17. Letters in Box 7, Smith Papers; Arkell, MacMillan, and Stewart, eds., *Aspen, Pitkin County, Colorado: Her Mines and Mineral Resources* (Aspen: Aspen Daily Leader, July 1, 1892), 25; "Gleanings," *RMN* (October 28, 1868): 4; "Pitkin," *RMN* (July 28, 1885): 6; Brown to Smith, April 8, 1887, Box 1 FF 19, Smith Papers; Arkell, MacMillan, and Stewart, *Aspen, Pitkin County, Colorado*, 19; "Deep Mining," *DT* (December 29, 1901): section 3, p. 8; Nichols to Smith, June 15, 1894, Box 3 FF 6, Smith Papers.

18. Smith to Moffat, March 22, 1893, Box 5 FF 10, Smith Papers.

19. Smith to Moffat, May 13, 1892, May 14, 1892; both in Box 5 FF 10, Smith Papers.

20. Robert L. Brown, *Colorado Ghost Towns: Past and Present* (Caldwell, ID: Caxton, 2002), 84–85; Flinn to Smith, December 21, 1890, December 31, 1890, January 9, 1891; all in Box 2 FF 8, Smith Papers.

21. Unaddressed, undated letter in Mining Interests–Colorado (Pitkin) FF, Smith Papers.

22. Unaddressed, undated letter, Smith Papers.

23. Arthur Nichols to Smith, June 15, 1894, Box 3 FF 6; D.R.C. Brown to Smith, September 3, 1891, Box 1 FF 19; both in Smith Papers; *Mines and Mining Men in Colorado*, 20–21.

24. Nichols to Smith, June 15, 1894; *Mines and Mining Men in Colorado*, 17.

25. *Carbonate Chronicle* (Leadville) (July 18, 1892), quoted in Griswold and Griswold, *History of Leadville and Lake County Colorado*, vol. 2, 2013; description of the Austin process in Stewart, *Thomas F. Walsh*, 36.

26. George Brown Tindall and David E. Shi, *America: A Narrative History*, 4th ed., vol. 2 (New York: W. W. Norton, 1996), 945, 950, 959.

27. *Evening Chronicle* (Leadville) (June 27, 1893), quoted in Griswold and Griswold, *History of Leadville and Lake County Colorado*, vol. 2, 2017.

28. Fanny Watson to Patrick O'Kane (quoting Cora), July 1893, Watson and O'Kane Family Letters.

29. "To-Day's Convention," *RMN* (July 11, 1893): 1; "Solid Front," *RMN* (July 12, 1893): 1.

30. "To the East," *RMN* (July 13, 1893): 1–2.

31. *Leadville Herald Democrat* (September 3, 1893), quoted in Griswold and Griswold, *History of Leadville and Lake County Colorado*, vol. 2, 2021; *Leadville Herald Democrat* (September 15, 1893), quoted in Griswold and Griswold, *History of Leadville and Lake County Colorado*, vol. 2, 2022.

Chapter 4: "I Handle All the Mining Business"

1. Elizabeth Jameson, *All That Glitters: Class, Conflict, and Community in Cripple Creek* (Urbana: University of Illinois Press, 1998), 21, 42–45; "Robert Womack Suffers a Stroke of Paralysis," *DR* (July 28, 1904): 7; "Victor Mines," *Rocky Mountain Herald* (Denver) (August 10, 1895): 1–2: Abbott, Leonard, and McComb, *Colorado*, 113–114.

2. "Victor Mines," *Rocky Mountain Herald* (August 10, 1895): 1–2; Aldrich to Smith, May 4, 1892, Box 1 FF 2, Smith Papers.

3. Agreement between Smith and Roudebush, June 15, 1892, Box 7 Anaconda Mines FF; Hoskins to Smith, 1894, Box 2 FF 2; Smith to George Tilden, undated but probably around December 7, 1895, BV 3; all in Smith Papers.

4. Smith to Tilden, undated but probably around December 7, 1895.

5. Jameson, *All That Glitters*, 53–54; Tindal and Shi, *America*, 872.

6. Duane A. Smith, "Chaffee: The Man behind the Name," *Colorado Central Magazine* 114 (August 2003): 19; "Daily News," *RMN* (November 12, 1873): 4.

7. Jameson, *All That Glitters*, 54; Thomas M. Patterson quoted in "The Long Strike at Cripple Creek: Complete History of the Miners' Struggle in the Great Gold Camp," *Great Divide* 11 (July 1894): 161.

8. Jameson, *All That Glitters*, 54.

9. Jameson, *All That Glitters*, 54–55; Lipsey, *Lives of James John Hagerman*, 158–159; Eben Smith Memorandum Book, Box 18, Smith Papers.

10. Jameson, *All That Glitters*, 55; Patterson, "Long Strike at Cripple Creek," 161.

11. Jameson, *All That Glitters*, 55–56.

12. Jameson, *All That Glitters*, 56–57; Patterson, "Long Strike at Cripple Creek," 161.

13. Parker to Smith, April 25, 1894, Box 3 FF 12, Smith Papers.

14. Smith to Hagerman, June 28, 1895, BV 2; Smith to Keith, June 30, 1894, BV 1; Smith to Hagerman, July 5, 1894, BV 1; all in Smith Papers.

15. Jameson, *All That Glitters*, 57–58; Lipsey, *Lives of James John Hagerman*, 176; "Petition from Mine Owners to El Paso County Board of Commissioners," May 17, 1894, quoted in Lipsey, *Lives of James John Hagerman*, 169–171; Patterson, "Long Strike at Cripple Creek," 161.

16. Jameson, *All That Glitters*, 58; Buell to Smith, March 9, 1894, Box 1 FF 22; Jordan to Smith, May 31, 1894, Box 2 FF 31; both in Smith Papers; Jordan, "Jerome Chaffee and Eben Smith," Smith MSS.

17. Lipsey, *Lives of James John Hagerman*, 180–181; Smith to Hagerman, May 29, 1894; Smith to Hagerman, May 30, 1894; both letters quoted in Lipsey, *Lives of James John Hagerman*, 185–187.

18. Telegram from Campion to Smith, June 1, 1894, Box 1 FF 27, Smith Papers; Jameson, *All That Glitters*, 58; Memo of Meeting of David H. Moffat, J. J. Hagerman, and Davis H. Waite, quoted at length in Lipsey, *Lives of James John Hagerman*, 191–209; "Agreement between David H. Moffat, J. J. Hagerman, and Davis H. Waite," June 4, 1894, Box 27092 FF 15, Davis H. Waite Papers, Colorado State Archives, Denver; Ballou to Smith, June 5, 1894, Box 1 FF 6; Watson to Smith, June 7, 1894, Box 3 FF 51; both in Smith Papers; Email from Anne McCarroll to David Forsyth, August 29, 2003. In this email, McCarroll wrote that her father, Louis O'Kane, always said that "old Joe Watson" would never carry a gun.

19. Jameson, *All That Glitters*, 58.

20. Smith to Trevorrow, July 9, 1894, BV 1; Reid to S. K. Adams, October 25, 1894, BV Victor 1; Rockwell to Smith, August 1, 1894, Box 3 FF 26; all in Smith Papers; Jameson, *All That Glitters*, 59.

21. Edward T. Bollinger, "Moffat Road Built for Less than Estimate," *Steamboat Pilot* (July 30, 1959): 6B; B. Travis Wright and Kate Wright, *Images of America: Rollins Pass* (Charleston, SC: Arcadia, 2018), 26; Mehls, *David H. Moffat, Jr.*, 137–139.

22. Smith to Keith, September 28, 1894, BV 1; W.B.S. of Pinkerton's Detective Agency to Smith, April 27, 1895, Box 3 FF 16; W.B.S. to Smith, April 27, 1895, Box 3 FF 16; all in Smith Papers; G.H.S. to Smith, June 5, 1895, Tritt Collection; W.B.S. to Smith, May 5, 1895; May 8, 1895; May 16, 1895; all in Box 3 FF 16, Smith Papers.

23. G.H.S. to Smith, May 29, 1895; May 31, 1895 (2); June 5, 1895; June 8, 1895; McParland to Smith, June 8, 1895; all in Tritt Collection.

24. Reid to J. W. Newell, February 8, 1894, Box 10 FF B; Eben Smith to Eben Smith, December 19, 1894, Box 10 FF H; Receipt from Tabor Real Estate Company, February 12, 1895, Box 8 FF T (2); Barnes to Smith, May 15, 1894, Box 1 FF 4; Colorado Gold Reduction Works Company Prospectus, 1894, Box 7 FF Colorado Gold Reduction Works Company (second quotation); Statement of Corporation of the Resurrection Gold Mining Company, May 16, 1894, Box 7 FF Resurrection Gold Mining Company; Resurrection Gold Mining Company Meeting Minutes, November 15, 1895, Box 7 FF Resurrection Gold Mining Company; all in Smith Papers.

25. Minutes of Board of Directors Meeting of the Victor Gold Mining Company, June 16, 1894, Box 13 FF M (1); Smith to Moffat, December 26, 1894, BV Victor 1; both in Smith Papers.

26. Receipt from George H. Hill, surgeon dentist, January 1, 1894, Box 8 FF H; Smith to Reid, January 21, 1895, Box 5 FF 11; both in Smith Papers.

27. Moffat to Smith, May 8, 1895, Box 13 FF M (2); Horton to Smith, July 22, 1895, Box 13 FF H (2); both in Smith Papers.

28. Statement of sale of Battle Mountain Gold Mining Company stock, n.d., Box 7 FF Battle Mountain Gold Mining Company; Lennox to Smith, June 25, 1895, Box 2 FF 40; Smith to Webb, February 10, 1903, Box 5 FF 19; all in Smith Papers.

29. Agreement between Baka-Contact Gold Mining Company and Eben Smith and David Moffat, September 1, 1894, Box 10 FF W, Smith Papers.

30. Smith to Foster, August 21, 1894, BV 1; Smith to Crawford (citing Crawford letter), September 27, 1894, Box 5 FF 10; Contract, November 10, 1894, Box 7A FF Mining–Foreign Investments (Mexico); all in Smith Papers.

31. Smith to Wood, August 24, 1894, BV 1; Morrison to Smith, June 5, 1894; July 2, 1894; August 8, 1894; August 11, 1894, all in Box 2 FF 53; all in Smith Papers.

32. Agreement between George E. Collins and Eben Smith, January 18, 1895; Lease between George E. Collins and Eben Smith, January 15, 1895, both in Box 7A FF Mining Interests–Colorado Chaffee County; Contract between Antlers-Park Regent Leasing Company and Eben Smith and David Moffat, February 16, 1895, Box 10 FF XYZ; Smith to Barnett, April 26, 1895, Box 10 FF Antlers-Park; Barnett to Stockholders of Antlers-Park, July 16, 1895, Box 10 FF Antlers-Park; Barnett to Reid, October 10, 1895, Box 10 FF Antlers-Park; all in Smith Papers.

33. Poole to Charles Martin, March 13, 1895, Box 10 FF P Antlers-Park; Jennings to Smith, June 6, 1895, Box 10 FF IJ Antlers-Park; both in Smith Papers.

34. Dean to Ray, January 23, 1895, Box 11 FF United Leasing D, Smith Papers.

35. Smith to Ray, June 18, 1895, BV 2; Bryant to Moffat, January 13, 1896, Box 11 FF United Leasing B; Lien filed by Thomas Armstrong against United Leasing Company, January 16, 1896, Box 11 FF United Leasing L; all in Smith Papers.

36. Bowers to Reid, January 17, 1896, Box 11 FF United Leasing B; Writ of Attachment filed by Thomas Armstrong against United Leasing Company, February 6, 1896, Box 11 FF United Leasing L; Barnie and Rue to Moffat, February 10, 1896, Box 11 FF United Leasing B; Armstrong to Moffat, February 25, 1896, Box 11 FF United Leasing A; Armstrong to Reid, February 14, 1896, Box 11 FF United Leasing A; all in Smith Papers.

37. Contract between Eben Smith and David Moffat and Joel Parker Whitney, October 22, 1895, Box 7 FF Raven Gold Mining Company, Smith Papers.

38. Smith to Knox, May 27, 1895, Box 5 BV 2; Norton to Smith, August 10, 1895; August 28, 1895; September 14, 1895; November 11, 1895; all in Box 3 FF 9, Smith Papers.

39. Smith to Argall, July 21, 1896, BV 4, Smith Papers; Jim Maxwell, "Built in 1902," in "Colorado Question Box," *RMN* (March 3, 1957): 51; "History of the Mine and Smelter Supply Company," WHGD.

40. Smith to Chalmers, June 26, 1895, BV 2; Smith to Argall, July 21, 1896, BV 4; both in Smith Papers.

41. Smith to Cone, July 5, 1894, BV 1; Cone to Smith, January 8, 1895; July 14, 1895; August 16, 1895; January 25, 1897; all in Box 1 FF 36, Smith Papers.

42. Application for accident insurance, February 3, 1897, Box 7A FF Smith, Eben Insurance (Accident Policy); Smith to Keith, June 18, 1895, BV 2; Smith to Ray, June 18, 1895, BV 2; all in Smith Papers.

43. Smith to Frank Sutherland, August 29, 1898, BV 7; Reid to Keith, February 27, 1895, BV Victor 1; Smith to Hendricks, August 24, 1894, BV 1; all in Smith Papers.

44. 1895 Annual Report of the Victor Gold Mining Company, Box 13 FF A (3); Smith to Reid, January 22, 1896; January 25, 1896, both in Box 5 FF 12; all in Smith Papers.

45. Smith to Cone, July 18, 1896, BV Victor 2; 1896 Annual Report of the Victor Gold Mining Company, Box 13 FF A (4); both in Smith Papers.

46. Smith to Cone, May 15, 1897, BV 5; Smith to Reid, 1897, Box 5 FF 13; both in Smith Papers.

47. Denver Union Water Company Bill, November 1, 1895, Box 8 FF D, Smith Papers; "The Eben Smith Residence," *DR* (June 8, 1896): 2; Notice from Smith to the Colorado Telephone Company, March 26, 1894, Box 8 FF D, Smith Papers.

48. Lake County, Colorado Marriage Records, Lake County Clerk and Recorders Office, Leadville, CO; Articles on Carnahan Shooting, 444–445, William M. Wise Scrapbook, WHGD; Receipt from Singer Sewing Machine Company, Box 8 FF S, Smith Papers; Robert Reid to Emily Smith, August 9, 1900, Box 5A FF S, Smith Papers; Arapahoe County Marriage Committee, *Marriages of Arapahoe County, Colorado, 1859–1901* (Denver: Colorado Genealogical Society, Inc., 1986), for Frank LeRoy Smith and Josie Bonita Hill; "C. L. Hill, Who Died on Coast, Noted Colorado Mining Man," *DT* (January 21, 1912): 6; "Secret Wedding for a Denver Millionaire," *DT* (December 7, 1909): 1, 7; "Judge Severs Bonds of Son's Old Inamorata: Allen Grants Divorce to Mrs. Frank Leroy Smith, Once Prospective Daughter-in-Law," *RMN* (November 5, 1907): 5; Eben Smith mausoleum, Fairmount Cemetery, Denver, CO.

49. Receipts from J. G. Kilpatrick Artistic Furniture and Upholstery, September 29, 1894; October 31, 1894, both in Box 8 FF K; Reid to Emily Smith, August 9, 1900, Box 5A FF S; all in Smith Papers; Cora Carnahan to Fanny O'Kane, January 25, 1899, Watson and O'Kane Family Letters.

50. Receipts from the Alta Market Company, May 1, 1894; June 1, 1894, both in Box 8 FF A; Smith to Reid, September 25, 1901, Box 5A FF S; Dines to Smith, January 12, 1903, Box 1 FF 45; Receipt from Bryan, Taylor and Company, Box 8 FF B; all in Smith Papers; Noel, ed., *Glory That Was Gold*, 266.

51. Receipt from H. M. Chamberlain, May 27, 1895, Box FF D; Receipt from J. H. Tummer, October 23, 1895, Box 8 FF T; Receipt from Thomas Chapman, florist, May 1, 1896, Box 8 FF B; all in Smith Papers.

Chapter 5: "Our Various Operations"

1. Smith to Reid, January 21, 1896, Box 5 FF 12; Reid to Postmaster, October 7, 1895, BV 3; Smith to Reid, January 13, 1896; January 21, 1896; January 17, 1896; January 25, 1896, all in Box 5 FF 12; all in Smith Papers.

2. Smith to Reid (citing the telegram), January 21, 1896, Box 5 FF 12, Smith Papers; Estate of Emily Smith, July 28, 1910, *Inheritance Tax Record Book Three*, City and County of Denver, 61; Smith to Johnson, March 25, 1896, BV 4, Smith Papers.

3. Telegram from Smith to Reid, February 4 5, 1896; Telegram from Smith to Reid, February 6, 1896, both in Box 5 FF 12, Smith Papers; "Death of Henry Smith," *Evening Chronicle* (Leadville) (February 5, 1896): 1; Eben Smith to Mark Smith, February 26, 1896; Reid to Bassett, March 2, 1896; Reid to Lyons, March 2, 1896, all in BV 4, Smith Papers; Smith, *Brief History and Complete Record of the American Descendants of John and Mary Smith*, 40–43, Tritt Collection.

4. Smith to Cone, December 13, 1897, BV 6; Smith to Charles A. Keith, July 22, 1894; Smith to H. L. Van Nostrand, July 23, 1894; Smith to Johnson, September 27, 1894, all in BV1; all in Smith Papers.

5. Limberg to Smith, March 31, 1896, Box 2 FF 41; Curnow to Smith, June 22, 1896, Box 1 FF 25, both in Smith Papers; "Daily News," *RMN* (January 17, 1870): 4; Smith to Curnow, July 7, 1896, BV 4; Curnow to Smith, July 24, 1896, Box 1 FF 25; Smith to Nichols, August 3, 1896, Box 5 FF 12; all in Smith Papers.

6. Smith to Ridgeway, March 31, 1897, BV 5; Smith to Sweet, January 6, 1896, BV 3; both in Smith Papers.

7. Griswold and Griswold, *History of Leadville and Lake County Colorado*, vol. 2, 2155–2156, 2159; *Carbonate Chronicle* (Leadville) (February 8, 1897), quoted in Griswold and Griswold, *History of Leadville and Lake County Colorado*, vol. 2, 2159. Cloud City was another name for Leadville.

8. Agreement between Eben Smith, A. V. Bohn, S. W. Mudd, and John F. Campion to found the Colorado Mines Association, June 12, 1896, Box 7A FF Colorado Mines Association, Smith Papers.

9. Griswold and Griswold, *History of Leadville and Lake County Colorado*, vol. 2, 2156, 2159.

10. Cornish to Smith, June 19, 1896, Box 1 FF 37, Smith Papers; Griswold and Griswold, *History of Leadville and Lake County Colorado*, vol. 2, 2159.

11. Tindall and Shi, *America*, 959–960.

12. Smith to Horton, July 30, 1896, BV 4, Smith Papers.

13. Smith to Horton (citing Horton's letter), July 30, 1896; Smith to Thomas, September 25, 1896; both in BV 4, Smith Papers.

14. Shelton to Smith, September 22, 1896, Box 3 FF 33; H. J. Higgins to Charles T. Carnahan, January 16, 1897, Box 5 FF 4; Shelton to Smith, October 2, 1896, Box 3 FF 3; all in Smith Papers; Griswold and Griswold, *History of Leadville and Lake County Colorado*, vol. 2, 2160, 2163, 2169.

15. Jordan to Smith, May 31, 1894; Telegram from Jeffery and Jordan to Smith, November 7, 1896, both in Box 2 FF 31; Reid to Campion, November 9, 1896, BV 5; all in Smith Papers; Listing for Nellie Smith and Charles Stuart Jeffrey [*sic*], California, County Marriages, 1850–1952, www.familysearch.org, last accessed April 29, 2017; Listing for Mrs. Nellie Jeffery, California, Oakland, Mountain View Cemetery Records; telegram from Jordan to Smith, November 14, 1896, Box 2 FF 31, Smith Papers; "Obituary for Nellie Jeffery"; Smith Family Plot, Mountain View Cemetery; Smith to Dorsey, November 18, 1896, BV 5, Smith Papers.

16. Smith to Parrish, November 28, 1896; Smith to Goad, November 23, 1896; Smith to Goad, November 23, 1896, all in BV 5; Agreement between E. H. Crawford and Eben Smith, David Moffat, Sylvester Smith, and Lafayette Campbell,

December 12, 1896, Box 7A FF Mining Interests-Agreements; all in Smith Papers; "New Rails to the Penrose," *RMN* (October 21, 1899): 10.

17. Smith to Reid, December 23, 1896, Box 5 FF 12, Smith Papers.

18. Smith to Reid, December 26, 1896, Box 5 FF 12, Smith Papers.

19. Cone to Smith, December 31, 1896; Reid to Smith, December 31, 1896; both in BV 5, Smith Papers.

20. Smith to Reid, December 26, 1897, Box 5 FF 12; Smith to Reid, January 3, 1897 (second quotation); January 6, 1897 (third quotation), both in Box 5 FF 13; all in Smith Papers.

21. Smith to Moffat, January 10, 1897 (quotation); January 13, 1897; both in Box 5 FF 13, Smith Papers.

22. Smith to Moffat, January 13, 1897; Smith to Reid, January 22, 1896, both in Box 5 FF 12, Smith Papers; Gressley, *West by East*, 6.

23. Smith to Canning, February 9, 1897; Smith to Cone, February 20, 1897; both in BV 5, Smith Papers; Griswold and Griswold, *History of Leadville and Lake County Colorado*, vol. 2, 2169–2170; Higgins to Carnahan, January 26, 1897, Box 5 FF 4; Limberg to Smith, April 11, 1897, Box 2 FF 41; both in Smith Papers.

24. Smith to Brevoort, March 8, 1897; Smith to Horton, September 25, 1896; Smith to Cone, March 17, 1897; all in BV Victor, Smith Papers.

25. Moffat to Smith, April 30, 1897, Box 2 FF 50, Smith Papers.

26. Dorsey to Smith, April 8, 1897, Box 1 FF 46; Searles to Whom It May Concern, May 19, 1897, Box 3 FF 20; Smith to Brevoort, May 5, 1897, BV 5; all in Smith Papers.

27. Smith to Cone, September 18, 1896, BV 4; May 15, 1897, BV 5; Smith to Moffat, May 15, 1897, Box 5 FF 13; Moffat to Smith, April 30, 1897, Box 2 FF 50; all in Smith Papers.

28. Smith to Reid, May 20, 1897; May 30, 1897; both in Box 5 FF 13, Smith Papers.

29. Smith to Reid, May 30, 1897; June 2, 1897, both in Box 5 FF 13; Reid to Smith, June 23, 1897, BV 5; all in Smith Papers.

30. Smith to Reid, June 13, 1897; June 18, 1897; Smith to Moffat, July 8, 1897; all in Box 5 FF 13, Smith Papers.

31. Smith to Reid, July 11, 1897; July 13, 1897; August 11, 1897; September 11, 1897; all in Box 5 FF 13, Smith Papers.

32. Griswold and Griswold, *History of Leadville and Lake County Colorado*, vol. 2, 2171, 2186, "New Rails to the Penrose," *RMN* (October 21, 1899): 10; Limberg to Smith, October 18, 1897, Box 2 FF 41; November 5, 1897, Box 2 FF 13; both in Smith Papers.

33. Receipts from the Bohm-Bristol Company, April 4, 1896; June 1, 1896; both in Box 8 FF B, Smith Papers. Numerous receipts similar to these two are scattered throughout the files in Boxes 8 and 9 of the Smith Papers. Eben Smith mausoleum,

Fairmount Cemetery; Frank Leonard Smith, World War I Draft Registration Card, www.familysearch.org, last accessed May 25, 2019.

34. Receipt from Robertson and Doll Carriage Company, August 1, 1895, Box 8 FF R; Receipt from A. T. Demarest and Company Carriage Builders, May 20, 1898, Box 8 FF A (2); Receipt from Bon Ton Shoeing Shop, March 31, 1899, Box 8 FF B (3); all in Smith Papers; "'Estamere,' Beautiful Seat of Eben Smith, at Palmer Lake," *DT* (October 20, 1901): 16 (quotation); "Eben Smith Buys Estmere [*sic*]," *DT* (July 2, 1898): 5; Frank L. Smith to Reid, July 29, 1898, Box 8 FF S (2); Receipt to Frank L. Smith, July 31, 1898, Box 8 FF N (2); both in Smith Papers. The house's name is now spelled Estemere, but when it was built and throughout the time the Smith family owned it, the name was spelled Estamere. The earliest use of the Estemere spelling was in 1907 according to Daniel W. Edwards and Roger W. Ward, *A Victorian Mansion in the Colorado Rockies: The Estemere Estate at Palmer Lake* (Privately published by the authors, 2012), xiv.

35. Eben Smith to Melvin Smith, January 15, 1898; February 28, 1898 (first quotation), both in BV 7; Jordan to Smith, May 10, 1898; June 13, 1898 (second quotation); all in Box 2 FF 27, Smith Papers.

36. J. S. McGinnis and Company to Smith and Moffat, June 12, 1898, Box 11 FF M Gold Knob Mining Company; J. J. Post, F. R. Power, T. A. Allen, Dr. H. E. Driscoll, and M. J. Driscoll to Reid, October 6, 1899, Box 11 FF PQ Gold Knob Mining Company; both in Smith Papers.

37. December letters, BV Victor, Smith Papers, for resumption of dividends; Betty Works to David Forsyth, February 15, 2004.

Chapter 6: "Not a Pound of Ore in Sight"

1. Brevoort to Smith, February 14, 1898; March 28, 1898; both in Box 1 FF 15, Smith Papers; Tindall and Shi, *America*, 979–980; Brevoort to Smith, November 1, 1900, Box 5A FF B, Smith Papers.

2. Agreement between Golden Circle Railroad Company and Eben Smith, 1898, Box 7A FF Golden Circle Railroad Company Statements, Smith Papers.

3. Smith, Entry in Excelsior Diary, July 27, 1898; July 28, 1898; both in Box 18, Smith Papers.

4. Cone to Smith, September 19, 1898, Box 1 FF 36; Smith to Cone, November 29, 1898, BV 7; Smith, Entry in Excelsior Diary, September 24, 1898, Box 18; all in Smith Papers.

5. Check from Eben Smith to Granite Gold Mining Company, July 2, 1898; Check from Eben Smith to Gold Knob Mining Company, July 14, 1898; both in Box 17, Smith Papers.

6. Cone to Smith, December 12, 1898, Box 1 FF 36; Reid to Sutherland, April 13, 1899, BV 8; Underwood to Cone, Box 4 FF 1; all in Smith Papers.

7. Bushnell to Smith, May 17, 1899, Box 1 FF 24, Smith Papers; Eben Smith mausoleum, Fairmount Cemetery.

8. Smith to Deweese, June 7, 1899, Box 5 FF 15, Smith Papers; Listing for Raymond and Grace Rippley, Colorado Springs, CO, US Bureau of the Census, *Twelfth Census of the United States, 1900* (Washington, DC: Government Printing Office).

9. Smith to Deweese, June 7, 1899, Box 5 FF 15; Berry to Smith, December 7, 1899, Box 1 FF 4; Unsigned to Berry, January 12, 1900, Box 5 FF 16; all in Smith Papers; Listing for Raymond R. and Grace R. Ripley, San Francisco, CA, US Bureau of the Census, *Thirteenth Census of the United States, 1910* (Washington, DC: Government Printing Office). The Ripleys did not divorce at the time Eben Smith wrote his letters, as they were married in the 1900 Census and still married in the 1910 Census.

10. Smith to Moffat, August 11, 1899, BV 8; Cone to Smith, November 15, 1898; May 1, 1899, both in Box 1 FF 36; Brevoort and Reid to Victor Shareholders, October 17, 1900, Box 14 FF O (5); all in Smith Papers.

11. Receipts from George R. Tritch Hardware Company, July 28, 1899; June 30, 1899; July 17, 1899; all in Box 8 FF Receipts 1898–1900, Smith Papers; John W. Pitts and Suzanne Pitts, *The Estemere Estate of Palmer Lake, 1887–1999* (Palmer Lake, CO: Palmer Lake Historical Society, 1999), 6; Smith to Moffat, August 11, 1899, BV 8; Moffat to Smith, August 15, 1899, Box 2 FF 50; Smith to M. H. Lindsey, September 11, 1899, BV 8; all in Smith Papers.

12. "New Rails to the Penrose," 10.

13. Osgood to Reid, November 14, 1899, Box 14 FF O (4); Smith to Wood, November 14, 1899, Box 14 FF W (4); both in Smith Papers.

14. Contract between Edward Thomas and Eben Smith, November 1, 1899, Box 7 FF Colorado-Sonora Mining and Milling Company; Smith to Wood, November 20, 1899, BV 8; both in Smith Papers.

15. Smith to Goad, November 27, 1899, BV 8, Smith Papers.

16. Certificate of Incorporation for the Denver-Galena Zinc Mining Company, December 29, 1899, Box 16 FF Denver-Galena Zinc Mining Company, Smith Papers.

17. Gold Knob Mining Company Distribution Book, January 1900, Oversize, Gold Knob Mining Company Distribution; Allen to Smith, January 20, 1900, Box 1 FF 1; both in Smith Papers.

18. Smith to Moffat, February 2, 1900; Reid to Smith, February 23, 1900; both in BV 8, Smith Papers.

19. Smith to Reid, February 8, 1900; February 11, 1900; both in Box 5 FF 16, Smith Papers.

20. Smith to Reid, February 17, 1900; February 27, 1900; March 27, 1900; all in Box 5 FF 16, Smith Papers.

21. Steven F. Mehls, "Success on the Mining Frontier: David H. Moffat and Eben Smith, a Case Study," *Essays in Colorado History* 1 (1983): 101; Smith to Reid (quotation), undated but probably March 1900, Box 5 FF 16; Smith to Reid, March 25, 1900; Whiteford to Smith, June 11, 1900, both Box 5A FF W; all in Smith Papers.

22. Smith to Reid, March 25, 1900, Box 5 FF 16; Smith to Reid, April 1, 1900 (second and third quotations); April 5, 1900, both in Box 5A FF S; all in Smith Papers.

23. Smith to Reid, April 20, 1900, Box 5A FF S, Smith Papers; Abbot, Leonard, and McComb, *Colorado*, 262–266, 377–378.

24. Smith to Reid, March 8, 1900; Smith to Wilfley, March 13, 1900; both in Box 5 FF 16, Smith Papers.

25. Smith to Moffat, June 13, 1900, BV 9; Swain to Smith, June 14, 1900, Box 3 FF 39; both in Smith Papers.

26. Complaint of David Moffat against the Victor Gold Mining Company, August 15, 1900, Box 14 FF M (5), Smith Papers.

27. Hurtado to Brevoort, March 29, 1900, Box 14 FF H (4), Smith Papers.

28. Complaint of David Moffat against the Victor Gold Mining Company, August 15, 1900, Box 14 FF M (5); Smith to Brevoort, September 7, 1900, BV 9; Smith to Moffat, September 24, 1900, BV 9; Dorsey to Smith and Moffat, November 10, 1900, Box 9 FF A; all in Smith Papers.

29. Receipt from Dorsey to Smith, November 15, 1901, Box 9 FF A; Brevoort and Reid to Victor Shareholders, October 17, 1900, Box 14 FF O (4); Brevoort to Reid, November 1, 1900, Box 5A FF B; all in Smith Papers.

30. Underwood to Smith, September 9, 1900; October 3, 1900 (referencing Smith's letter); both in Box 5A FF UV, Smith Papers.

31. Smith to Wood, November 16, 1900, BV 9, Smith Papers.

32. Smith to Bushnell, June 29, 1900; Smith to Richardson, Hill and Company, September 4, 1900; both in BV 9, Smith Papers.

33. Contract, November 5, 1900, Box 5A FF R; Check from Smith to Independence Consolidated Gold Mining Company, August 9, 1900, Box 17; Smith to Reid, December 2, 1900, Box 5A FF S; all in Smith Papers.

34. Baker to Smith, February 27, 1900, Box 1 FF 4; Ritchie to Smith, April 22, 1900; May 2, 1900, both in Box 5A FF R; Woodbury to Smith, September 29, 1900; September 7, 1900; October 4, 1900, all in Box 5A FF W; all in Smith Papers.

35. "New Bonanza Mine in San Juan County," *Denver Times* (January 24, 1901): 9; "San Juan County," *Plaindealer* (Ouray) (January 25, 1901): 8; "'Estamere,' Beautiful Seat of Eben Smith," 6.

36. Carnahan to Smith, February 4, 1901, Box 1 FF 30; February 5, 1901 (quotation); March 5, 1902; both in Box 1 FF 31; all in Smith Papers.

37. Reid to Smith, April 15, 1901, BV 9, Smith Papers.

38. Reid to Smith, April 22, 1901, BV 9, Smith Papers.

39. "Geologist in Leadville," *DT* (July 12, 1901): 8.

40. "Eben Smith Brought First Stamp Mill into Colorado," *DT* (November 6, 1906): 2.

41. Carl Ubbelohde, Maxine Benson, and Duane A. Smith, *A Colorado History*, 7th ed. (Boulder: Pruett, 1995), 239–243.

42. Ubbelohde, Benson, and Smith, *Colorado History*, 268; "Ready To Retire; Eben Smith Proposes to Withdraw from Active Life," *DT* (June 2, 1901): 2.

Chapter 7: "I Cannot Pass a Good Mining Scheme By"

1. First National Bank of Denver to Smith, April 29, 1901, Box 5A FF (F); Smith to Welsh, June 1, 1901, BV 9; both in Smith Papers; "Purchased a Residence," *Los Angeles Herald* (May 11, 1901): 7.

2. Receipt from Eckstrom Wallpaper Company to Smith, July 23, 1901, Box 9 FF A (2); Munsell to Smith, August 3, 1901; August 19, 1901, both in Box 5A FF M; Welsh to Smith, August 6, 1901; August 14, 1901, both in Box 5A FF W; all in Smith Papers; Photographs of Eben Smith's house in "Smith House L.A. Photo Album," Tritt Collection; "Multi-Millionaire Eben Smith Dead," *Los Angeles Times* (November 6, 1906): part 2, p. 1.

3. "'Estamere,' Beautiful Seat of Eben Smith," 6.

4. "'Estamere,' Beautiful Seat of Eben Smith," 6.

5. "Ready to Retire; Eben Smith Proposes to Withdraw from Active Life," 2.

6. Smith to Buell (referencing Buell's letter), June 14, 1901, BV 9, Smith Papers.

7. Smith to Estes, August 1, 1901, BV 9; Mater to Smith, August 10, 1901, Box 2 FF 47; both in Smith Papers; "Denver Men Sell Their Utah Copper Property," *DT* (August 2, 1901): 11; "History of the Mine and Smelter Supply Company," Denver Public Library; "Reorganization of Big Machinery House," *DT* (May 16, 1901): 11; Jim Maxwell, "Built in 1902," in "Colorado Question Box," 51.

8. "Building and Real Estate Advices," *DT* (December 18, 1901): 9 (quotation); Smith to Carnahan, November 29, 1901, Box 1 FF 30, Smith Papers; Fanny O'Kane to Patrick O'Kane, September 8, 1901, Watson and O'Kane Family Letters; Receipt from Belle-Vernon Farms Company to Smith, December 1, 1904, Box 9 FF B (2), Smith Papers.

9. "Funeral Rites Set Friday for Denverite," *RMN* (August 15, 1963): 38; Carnahan to Smith, February 26, 1902; March 5, 1902; April 5, 1902; all in Box 1 FF 31, Smith Papers.

10. Cary to Smith, November 1, 1901, Box 5A FF C, Smith Papers; "Garfield Mine," *DT* (December 29, 1901): section 3, p. 5.

11. Receipt from Robert Reid, November 15, 1901, Box 9 FF A; Smith to Woodbury, December 10, 1901, Box 5 FF 17; Smith to Dines, December 18, 1901, Box 5 FF 17; Bond for a Deed between N. S. and Emma Berray, and Claude and Adele Batailleur and, Eben Smith and W. W. Swetland, March 10, 1902, Box 7A FF Mining Interests–Arizona; Sutherland to Smith, June 23, 1902, Box 3 FF 37; all in Smith Papers; "Eben Smith Leases Two Copper Claims, Arizona Globe District," *DT* (March 28, 1902): 13.

12. Thompson to Smith (referencing Smith's letter), April 30, 1902, Box 3 FF 44, Smith Papers; Smith, *Brief History and Complete Record of the American Descendants of John and Mary Smith*, 38, Tritt Collection; Smith to Thompson, July 9, 1902, BV 10, Smith Papers.

13. Agreement between Charles Mater and Eben Smith, May 28, 1902, Box 7A FF Mining Interests–Arizona; Mater to Smith, June 10, 1902, Box 2 FF 47; Smith to McCoy, June 11, 1902, BV 10; all in Smith Papers.

14. Memo of Meeting of Board of Directors of Granite Gold Mining Company, July 11, 1920, Box 7 FF Granite Gold Mining Company, Smith Papers; "Resurrection Mill for Low Grade Sulphides," *DT* (July 12, 1902): 9; Carnahan to Smith, January 22, 1903, Box 1 FF 31, Smith Papers.

15. "Report Gratifying; Extensive Development Done on Independence Consolidated Property," *DT* (August 8, 1902): 11 (referencing annual report).

16. Receipts from the Quentin Investment Company, August 20, 1902; August 22, 1902; August 23, 1902; August 25, 1902; August 26, 1920; Box 17A FF Investments–Quentin Investment Company, Smith Papers.

17. "To Spend $75,000 for a Home: Eben Smith Negotiating for a Site Facing City Park," *DT* (August 21, 1902): 1; "Building and Real Estate Advices," *DT* (December 18, 1901): 9.

18. Smith to McCoy, September 2, 1902, BV 10, Smith Papers.

19. Organization Meeting Minutes of the Utah Ice and Cold Storage Company, September 3, 1902, Box 7A FF Utah Ice and Cold Storage Company; Agreement between Gillespie and Smith, October 1902, Box 7 FF Guiana American Gold Company; both in Smith Papers.

20. The years of Dunham's association with Eben Smith are shown by the letters from Dunham to Smith in Boxes 1 FFs 47–50 and Box 2 FF 1 of the Smith Papers.

21. Dunham to Smith, November 21, 1902, Box 1 FF 48, Smith Papers.

22. Dunham to Smith, November 21, 1902, Box 1 FF 48; Dines to Smith, November 5, 1902, Box 1 FF 45; both in Smith Papers.

23. Dines to Smith, January 12, 1903; July 21, 1903; both in Box 1 FF 45, Smith Papers.

24. Moffat to Smith, December 20, 1902, Box 2 FF 50, Smith Papers.

25. Dines to Smith, January 12, 1903, Box 1 FF 45; Carnahan to Smith, January 17, 1903, Box 1 FF 31, both in Smith Papers.

26. Carnahan to Smith, January 17, 1903.

27. Eben Smith to Frank L. Smith, January 21, 1903, Box 5 FF 19; Hughes to Smith, August 17, 1903; Hughes to Dunham, August 17, 1903, both in Box 2 FF 24; all in Smith Papers.

28. Moffat to Smith, December 20, 1902; Power of Attorney, Eben Smith to Frank L. Smith, November 7, 1903, all in Box 7A FF Power of Attorney to Frank L. Smith, Smith Papers.

29. Smith to Lessig, September 2, 1902; Smith to Thompson (referencing Thompson's letter), September 2, 1902, both in BV 10; Thompson to Smith, February 5, 1903; December 9, 1902; June 5, 1903, all in Box 3 FF 44; all in Smith Papers.

30. Check from Smith to the Murray M. Harris Organ Company, November 20, 1902, Box 17; Smith to Armstrong and New, November 24, 1902, Box 5 FF 17; both in Smith Papers.

31. Smith to Carper, February 10, 1903, Box 5 File Folder 19; Carper to Smith, January 27, 1903; February 4, 1903, both in Box 1 FF 27; all in Smith Papers.

Chapter 8: "I Feel I Am in Hell's Hole"

1. Smith to Mater, September 2, 1902, BV 10; Mater to Smith, May 4, 1903, Box 2 FF 47; Smith to Central Trust Company, June 3, 1903; Smith to Dines, June 3, 1903; Smith to Evans, July 17, 1903; Smith to Farish, July 17, 1903; Smith to Hornblower and Weeks, July 17, 1903, all in BV 10; all in Smith Papers.

2. Dunham to Smith, August 5, 1903, Box 1 FF 49; Smith to Dunham, August 18, 1903, BV 11; both in Smith Papers.

3. Smith to Dunham, July 16, 1903; Smith to Reid, August 4, 1903; both in BV 10, Smith Papers.

4. Smith to Reid, August 4, 1903, BV 10; Ramsay to Smith, August 12, 1903, Box 3 FF 23; both in Smith Papers.

5. Smith to New, July 24, 1903, BV 10, Smith Papers.

6. "House Is Set on Fire to Hide a Burglary," *Los Angeles Herald* (April 23, 1903): 9; "Governess Arrested and Jewels Recovered," *San Francisco Call* (April 24, 1903): 13; "Larceny and Arson," *Los Angeles Herald* (April 24, 1903): 14.

7. Wise Scrapbook, 444–445; Smith to Dunham, undated but in August 1903, BV 11, Smith Papers.

8. Carnahan to Smith (referencing Emily Smith letter), August 21, 1903, Box 1 FF 31, Smith Papers; Wise Scrapbook, 451, 459–460 (includes newspaper articles); "Father-in-Law to Heal Breach," *Los Angeles Herald* (March 4, 1904): 3. The "Events of the World Epitomized" article on page 1 of the January 16, 1904, edition of the *Sausalito* (CA) *News* says Carnahan was sentenced to one day in jail and ordered to pay a $500 fine for the shooting, but the March 4, 1904, article in the *Los Angeles Herald* makes no mention of this.

9. Ramsay to Smith, August 27, 1903; Ramsay to Smith, September 3, 1903; both in Box 3 FF 21, Smith Papers.

10. Smith to Ramsay, September 7, 1903, BV 11; Ramsay to Smith, September 19, 1903, Box 3 FF 21; Howe to Smith, December 7, 1903, Box 2 FF 23; Smith to Ramsay, February 6, 1904, Box 5 FF 20; all in Smith Papers.

11. New to Smith, November 1, 1903, Box 3 FF 2, Smith Papers.

12. Check from Smith to Pacific Wireless, January 4, 1904, Box 17; Smith to New, January 10, 1904, Box 5 FF 20; Agreement between the Los Angeles Art Organ Company and Eben and Emily Smith, January 1904, Box 7A FF Business Interests–Los Angeles Art Organ Company; Smith to Ramsay, February 24, 1904, Box 5 FF 20; Smith to Ramsay, March 5, 1904, Box 3 FF 21; all in Smith Papers.

13. Helen Lukens Jones, "The Greatest of Pipe Organs," *Scientific American* 90 (April 23, 1904): 328–329.

14. Check from Smith to Los Angeles Art Organ Company, April 26, 1904; Check from Smith to Los Angeles Art Organ Company, May 18, 1904, both in Box 17; Smith to New, March 8, 1904, Box 5 FF 20; all in Smith Papers.

15. Brook to Smith, July 3, 1905; April 17, 1906; both in Box 1 FF 18, Smith Papers.

16. Checks from Smith to Los Angeles Art Organ Company, July 25, 1904; August 19, 1904, both in Box 17; Amounts Due Eben Smith from the Los Angeles Art Organ Company, September 2, 1904, Box 7A FF Business Interests–Los Angeles Art Organ Company; all in Smith Papers.

17. Check from Smith to Pacific Wireless Telegraph Company, October 4, 1904, Box 17; Smith to New, January 25, 1905, BV 12; both in Smith Papers.

18. "E. Smith Suffers Stroke: Paralysis Attacks Colorado Millionaire," *DP* (September 5, 1904): 2; Limberg to Smith, September 24, 1904, Box 2 FF 41, Smith Papers.

19. Moffat to Smith, March 26, 1903, Box 2 FF 50; Mater to Smith, April 19, 1904, Box 2 FF 47, both in Smith Papers; George G. Suggs Jr., *Colorado's War on Militant Unionism: James H. Peabody and the Western Federation of Miners* (Norman: University of Oklahoma Press, 1991), 81; Smith-Denison Lease Vouchers, 1904–1905–1906, Box 16, Smith Papers.

20. Thompson to Smith, January 7, 1904; April 19, 1904; May 4, 1904, all in Box 3 FF 44; Lapham to Lemuel Smith, undated, BV 9; Report of Stockholders Meeting of Joseph Gibbons Consolidated Mining and Milling Company, July 15, 1904, Box 7 FF Joseph Gibbons Consolidated Mining and Milling Company; all in Smith Papers.

21. Receipt from V. R. Bowers and Sons, November 4, 1904, Box 9 FF B (2); Smith to Lapham, November 13, 1904, Box 5 FF 20; Smith to Mater, January 8, 1905, Box 2 FF 47; all in Smith Papers.

22. Smith to New, January 25, 1905; February 25, 1905; February 21, 1905 (referencing the New letter), all in BV 12; April 17, 1906, BV 13; all in Smith Papers.

23. Phillips to Smith, February 22, 1905, Box 3 FF 15, Smith Papers.

24. Agreement between Spaulding and Smith, February 14, 1905, Box 7A FF Business Interests–Los Angeles Art Organ Company; Fleming to Smith, June 29, 1905, Box 2 FF 7; Brook to Smith, December 28, 1905, Box 1 FF 18; all in Smith Papers.

25. Smith to Lash, August 7, 1905 (referencing Lash's letter); Smith to Wolcott, November 22, 1905; both in BV 12, Smith Papers.

26. Russell to Smith, June 26, 1905, Box 3 FF 29; As president of Bakersfield and Ventura Railway, January 8, 1906, Box 1 FF 5; both in Smith Papers.

27. Lapham to Smith, July 13, 1905; Eben Smith to Frank L. Smith, August 9, 1905, both in BV 12; Smith to Sutherland, March 9, 1904, Box 5 FF 20; all in Smith Papers.

28. Smith to New, November 6, 1905, BV 12; Check to Los Angeles Art Organ Company from Smith, November 3, 1905, Box 17; Kislingbury to Smith, December 16, 1905, Box 2 FF 36; Carnahan to Smith, December 13, 1905, Box 1 FF 31; all in Smith Papers.

29. Smith to Ruple, December 11, 1905, BV 12, Smith Papers.

30. Dines to Smith, February 3, 1904, Box 1 FF 45; Telegram from Eben Smith to Emily Smith, June 2, 1904, Box 5 FF 20; both in Smith Papers.

31. Receipt from Felker Automobile Company to Smith, September 12, 1905, Box 9 FF 4; Lapham to Frank L. Smith, October 20, 1905, BV 12; Receipt from Westlake Stables Company to Smith, March 1, 1905, Box 9 FF W(3); Smith to Western Motor Car Company, January 25, 1906, Box 5 FF 22; Receipt from Western Motor Car Company, February 23, 1906, Box 9 FF W (4); all in Smith Papers.

Chapter 9: "The Dean of Mining in Colorado"

1. Smith to Brook, January 2, 1906, BV 12; Smith to Osborn, January 6, 1906, Box 5 FF 31 (referencing Osborn's letter); Limberg to Smith, January 3, 1906, Box 2 FF 41; all in Smith Papers.

2. Telegram from Emily Smith to Cora Carnahan, January 25, 1906, BV 12, Smith Papers; "Multi-Millionaire Eben Smith Dead," part 2, p. 1; Telegram from

Lapham to Frank L. Smith, January 26, 1906, BV 12; Kislingbury to Smith, January 26, 1906, Box 2 FF 36; all in Smith Papers; Caricature of Frank L. Smith in B. S. White, *Representative Men of the West in Caricature* (Denver: *American Cartoonist Magazine*, 1904), 297.

3. Lapham to Brook, February 8, 1906; Lapham to Coffin, February 17, 1906; Frank L. Smith to Ruple, February 19, 1906; Frank L. Smith to Mason, March 22, 1906, all in BV 12; Check from Smith to Electrolian Organ Company, March 5, 1906; March 24, 1906; April 18, 1906, all in Box 17; all in Smith Papers.

4. Charles F. Mater Jr. to Smith, March 8, 1906, Box 2 FF 47, Smith Papers.

5. Charles F. Mater to Smith (referencing Smith's letter), March 15, 1906, Box 2 FF 47, Smith Papers.

6. Charles F. Mater to Smith, March 15, 1906; Telegram from Charles F. Mater to Smith, August 6, 1906; both in Box 2 FF 47, Smith Papers.

7. On Frank being in Seattle, April 16, 1906; Smith to New, April 17, 1906; both in BV 13, Smith Papers.

8. Brook to Smith, April 17, 1906; May 8, 1906; June 7, 1906 (quotation), all in Box 1 FF 18; Smith to Busch, May 31, 1906; Smith to Brook, June 28, 1906, both in BV 13; all in Smith Papers.

9. "Multi-Millionaire Eben Smith Dead," part 2, p. 1; Receipt from Susie de Luna to Smith, June 23, 1903, Box 9 FF D (2); Smith to Cora Carnahan, May 18, 1906, BV 13; Telegram from Hersey to Smith, July 10, 1906, Box 2 FF 14; Telegram from Smith to Hersey, July 10, 1906; Smith to Campion, September 1, 1906, both in BV 13; all in Smith Papers.

10. Lapham to Smith, October 27, 1906; Lapham to Smith, November 2, 1906; both in BV 13, Smith Papers; "Broken Is the Golden Bowl, the Spirit Flown Forever," 1.

11. Telegram from Lapham to Emily Smith, November 6, 1906, BV 13, Smith Papers; "Broken Is the Golden Bowl, the Spirit Flown Forever," 1; "Death of Eben Smith," *Breckenridge Bulletin* (November 10, 1906), Eben Smith Clipping File; "Eben Smith Brought First Stamp Mill into Colorado," 2; "Eben Smith Dead," unknown source and date, Eben Smith Clipping File; "Eben Smith Dead at Ripe Old Age," *DR* (November 6, 1906): 1, 8; "Eben Smith, Pioneer and Mining Man, Dies," *DP* (November 6, 1906): 4; "Old Comrades at His Bier," *DT* (November 8, 1906): 2; "Pioneers Attend the Funeral of Eben Smith," *DR* (November 9, 1906): 7.

12. "Son of Eben Smith Cut Off with a Small Annuity," *DT* (November 15, 1906): 1, 9; "Son of Eben Smith Can Earn Big Sum by Mending His Ways," *DR* (November 16, 1906): 12; "$25,000 for Son if He Mends His Ways," *RMN* (November 16, 1902): 2; "Eben Smith's Son in Leadville, Working as Engineer, Faces Question," *DP* (November 16, 1906): 9.

13. Memo of ownership of Moffat Estate Company, n.d.; Vanderwilt and Parker to J. Churchill Owen, April 28, 1959; Schedule of Properties of the Moffat Estate Company in the Leadville District, Lake County, Colorado, all in Box 17 FF 24; Mining Deed, 1942, Box 13 FF 27; all in Moffat Estate Company Papers, WHGD.

14. "$50,000 for Lemuel Smith," *DT* (November 28, 1906): 1–2; Leadville City Directory, 1909; "Smith's Death Shows Unusual History of Life: Fatal Malady of Son of Millionaire Is Unexplained," *DP* (April 22, 1918): 4.

15. "$50,000 for Lemuel Smith," 1–2; "Eben Smith's Estate Worth a Million," *DT* (March 26, 1907): 4; "Divorce Divides the Eben Smith Millions: Heir of Colorado Mining Man Charged with Cruelty in Wife's Complaint," *DP* (November 4, 1907): 1, 4.

16. "District Attorney Denounces Defendant," *Los Angeles Herald* (November 17, 1906): 12; "Major Russell Answers Lawler," *Los Angeles Herald* (November 18, 1906): 8; "Seeks to Prevent Sale of a Strip of Railroad," *Los Angeles Herald* (December 2, 1906): 32; "Railroad Will Be Sold Today," *Los Angeles Herald* (December 21, 1906): 3; "Estate Buys Railroad," *San Bernardino Sun* (December 26, 1906): 12; "Incorporate New Railroad Company," *San Francisco Call* (August 1, 1908): 10; "New Incorporations," *Los Angeles Herald* (September 21, 1910): 8;

17. "Eben Smith's Estate Worth a Million," 4; "Sale on West Seventh," *Los Angeles Herald* (April 17, 1907): 12; "Monster Apartment Building Planned," *Los Angeles Herald* (September 12, 1909): 2; "Tender Orchids to Mayor Harper," *Los Angeles Herald* (February 24, 1907): 7; "Widow of Pioneer Mining Man Dead after Brief Illness," 16.

18. "Immense Alimony Allowed by Court: Son of Late Eben Smith Is Ordered to Pay Fortune," *DR* (November 5, 1907): 15; "Judge Severs Bonds of Son's Old Inamorata: Allen Grants Divorce to Mrs. Frank Leroy Smith, Once Prospective Daughter-in-Law," 5; "Divorce Court Ends the Smiths' Troubles: Wealthy Couple Parted," *DT* (November 7, 1907): 1, 3; "Divorce Divides the Eben Smith Millions," 1, 4.

19. "Secret Wedding for a Denver Millionaire," *DT* (December 7, 1909): 1, 7; "Widow of Rich Suicide to Wed Lawyer Who Settled Estate," *DP* (August 10, 1911): 1.

20. "Secret Wedding for a Denver Millionaire," 1, 7 (quotations); "Widow of Pioneer Mining Man Dead after Brief Illness," 16; "Death Claims Aged Widow of Late Eben Smith," 3; "Pioneer Woman Will Be Laid to Rest Today," 12; Frances Watson to Fanny O'Kane, December 26, 1909, Watson and O'Kane Family Letters; Eben Smith mausoleum, Fairmount Cemetery.

21. Josephine Howland to Fanny Watson O'Kane (citing Cora), December 26, 1909, Watson and O'Kane Family Letters; Estate of Emily Smith, *Inheritance Tax Record Book Three*, 61.

22. "Is Found with Bullet Wound in His Chest," *DR* (May 7, 1910): 1–2; "Frank Smith Shoots Self; Near Death," *DP* (May 7, 1910): 1 (quotation); "Frank L. Smith

Dies of Bullet Wound and Inquest Is Demanded," *DP* (May 9, 1910): 1–2; "Hold Smith Inquest Today; Wound Fatal," *RMN* (May 10, 1910): 2.

23. "Is Found with Bullet Wound in His Chest," *DR* (May 7, 1910): 1–2; "Frank Smith Shoots Self; Near Death," *DP* (May 7, 1910): 1 (second quotation); "Frank L. Smith Dies of Bullet Wound and Inquest Is Demanded," *DP* (May 9, 1910): 1–2; "Hold Smith Inquest Today; Wound Fatal," *RMN* (May 10, 1910): 2; "Frank Smith Left Estate of $550,000," *DP* (May 23, 1910): 1.

24. "Decide Shooting Was Intentional," *DR* (May 11, 1910): 2 (quotation); "Frank L. Smith Dies of Bullet Wound and Inquest Is Demanded"; Betty Works to David Forsyth, February 5, 2004.

25. "Widow of Rich Suicide to Wed Lawyer Who Settled Estate," 1; "Two Wives Claim Big Smith Estate," *Los Angeles Herald* (June 11, 1910): 4; Frances Watson to Fanny Watson O'Kane (citing Emily), October 10, 1909; Josephine Howland to Fanny Watson O'Kane, undated postcard, both in Watson and O'Kane Family Letters; Richard Wupperman to David Forsyth, April 10, 2004; "Worthington, Lula E.," *Sunday Star* (September 3, 1950): A-16.

26. "David H. Moffat Is Dead," 1, 6.

27. Incorporation of Smith-Moffat Mines Company, July 1, 1912, Box 17 FF 24; Articles of Incorporation, March 1912, Box 14 FF 7; Moffat Estate Company, n.d., Box 17 FF 24; Note from Richard Brown about Cora Carnahan's shares, undated, Box 16 FF 2; Report on the Victor-Isabella Project, August 30, 1936, Box 13 FF 40; all in Moffat Estate Company Papers.

28. Owen to Dwight, February 15, 1972, Box 16 FF 4, Moffat Estate Company Papers.

29. "Second Divorce to Woman Once Wife of Frank L. Smith," *DP* (February 18, 1912): 7; "Mrs. Hughes Gets Divorce Decree," *DT* (February 18, 1912): 23; "Second Divorce Secured in the Same Division," *DR* (February 18, 1912): 10; Betty Works to David Forsyth, February 5, 2004; "Mrs. Carnahan, Leader in Society, Granted Divorce," *DP* (April 4, 1915): section 1, p. 1; "Court Gives Carnahan's Wife Decree," *RMN* (April 4, 1915): 1; Listing for Thomas A. Costello, Denver City Directory, 1917; Listing for Thomas A. Costello in Denver, CO, US Bureau of the Census, *Fourteenth Census of the United States, 1920* (Washington, DC: Government Printing Office); "Eben L. Smith Killed in Fight before Sudan," *DT* (December 5, 1918): 3; Listings for Charles T. Carnahan at gencircles.com.

30. "Smith's Death Shows Unusual History of Life," *DP* (April 22, 1918): 4; Barbara Landweer Smith, telephone interview by the author, July 16, 2007; Smith Family Plot, Mountain View Cemetery; Listing for Nellie M. Smith in the California Death Index 1905–1939, www.familysearch.org, last accessed April 29, 2017.

31. "Margaret Onderdonk and Eben Smith, Jr., Elope," *DP* (July 8, 1914): 1; "Eben L. Smith Killed in Fight before Sudan," 3; "Eben Smith's $500,000 Will Go to Widow," *DT* (January 3, 1919): 11.

32. "Rich Ex-Denver Man Dies in Crash; M. H. Smith Is Killed in Coast City," *DT* (June 28, 1921): 1–2; Jim Maxwell, "Built in 1902," in "Colorado Question Box," 51; Listing for William E. Smith, born July 4, 1921, www.californiabirthindex.org, last accessed April 27, 2017. William Eben Smith died on October 16, 1973. Virginia Smith later married Henry Gerke, and he died on October 19, 1979. Both deaths are in the California Death Index 1940–1997, www.familysearch.org, last accessed April 27, 2017.

33. Corinne P. Sze, "The Frank Leonard Smith House," *Bulletin of the Historic Santa Fe Foundation* 20 (November 1992): 2; "Last Male Descendant of Eben Smith Succumbs Here to Sleeping Sickness; Death of Frank Leonard Smith, Grandson of Millionaire Mining Man, Is Third in Family within Last Few Years," *DP* (April 4, 1924): 1, 3; Betty Works to David Forsyth, February 15, 2004; *Margaretta Smith Hickey et al v. the Estate of Emily Smith and Cora I. Costello*, case number 11628, Supreme Court of Colorado, Colorado State Archives, Denver.

34. "Harold S. Carnahan Ends Life in California Sanitarium; Was Melancholia Patient for Years," *DP* (November 10, 1924): 1.

35. Minutes of Board Meeting, October 4, 1929, Box 14 FF 7; Deed, July 5, 1933, Box 13 FF 41; both in Moffat Estate Company Papers; "Mrs. Marie B. Carnahan," *Asheville* (North Carolina) *Citizen* (October 9, 1935): unknown page, copy provided by Asheville-Buncombe Library System; Charles and Marie Carnahan, Graves, Section S, lot 31, Riverside Cemetery, Asheville, NC.

36. Sze, "Frank Leonard Smith House," 2; "Obituary for Josephine Hill Smith," *RMN* (January 6, 1946): 47; Betty Works to David Forsyth, February 5, 2004.

37. Listing for Cora I. Costella (misspelling in original), New York City, NY, *Fourteenth Census of the United States, 1920*; Thomas A. Costello, Passport Application 364517, January 8, 1924, familysearch.org; Listing for Cora I. Costells (misspelling in original), Osborne, AZ, US Bureau of the Census, *Fifteenth Census of the United States, 1930* (Washington, DC: Government Printing Office); Listing for Mrs. Cora I. Costello, Denver City Directory, 1925; Listing for Mrs. Cora I. Costello, Denver City Directory, 1926; Listing for Mrs. Cora I. Costello, Denver City Directory, 1928; Listing for Thomas A. Costello, Los Angeles, CA, *Fifteenth Census of the United States, 1930*; Obituary for Thomas A. Costello, *Los Angeles Times* (July 17, 1968): part 3, p. 15; Listing for Mrs. Cora I. Costello, Denver City Directory, 1930; Noel, *Glory That Was Gold*, 265–266; Betty Works to David Forsyth, February 5, 2004. Dating Cora's divorce from Thomas Costello is difficult because she lists herself as divorced in

the 1920 Census, applied with Costello for a passport in 1924, and then listed herself as widowed in 1928. No record of their divorce has been found.

38. "Mrs. Cora Costello, State Pioneer, Dies," *DP* (January 22, 1956): 11a; "Cora I. Costello," *RMN* (January 18, 1956): 63. The comment from Louis O'Kane on Cora being unhappy is from an email from Anne McCarroll to David Forsyth, August 29, 2003. "Doris Van Schaack," *DP* (November 19, 1957): 13.

39. Smith-Moffat Mines to Doris Van Schaack, Emily Wilson, and the Moffat Estate, April 30, 1957, Box 16 FF 2; E. W. Anderson Retirement Memo, June 30, 1958, Box 17 FF 24; Mining Deed for Resurrection Mine, 1942, Box 13 FF 27; Vanderwilt and Parker to Owen, April 28, 1959, Box 17 FF 24; Deed, October 7, 1959, Box 13 FF 40; Max Bower to Smith-Moffat Mines Company, February 27, 1964, Box 16 FF 4; Balance Sheets 1 and 12, December 31, 1959, Box 17 FF 24; RGW to Owen, May 7, 1963, Box 16 FF 4; F. G. Moffat to Henry Hunington, March 26, 1903, Box 14 FF 2; all in Moffat Estate Company Papers.

40. "Funeral Rites Set Friday for Denverite," 38; "Emily Wilson," *DP* (August 15, 1963): 51; Smith, telephone interview by the author; J. Churchill Owen to George H.P. Dwight, February 25, 1972, Box 16 FF 4, Moffat Estate Company Papers.

41. Abbott, Leonard, and McComb, *Colorado*, 50–51; Gressley, *West by East*, 6.

42. Nearly all of the obituaries written about Eben Smith make the point that he was a driving force behind the development of mining in Colorado; "Jerome B. Chaffee's Death," 1; "Eben Smith, Pioneer and Mining Man Dies," *DP* (November 6, 1906): 4; Smith, *Trail of Gold and Silver*, 89.

43. "Eben Smith Dead at Ripe Old Age," 1, 8; "Eben Smith, Pioneer and Mining Man Dies," 4; "Eben Smith Brought First Stamp Mill into Colorado," 2.

44. "Eben Smith, Pioneer and Mining Man Dies," 4.

45. Smith, *Trail of Gold and Silver*, 228–229; "Emergency Response to August 2015 Release from Gold King Mine," https://www.epa.gov/goldkingmine, last accessed February 17, 2020.

46. Smith, *Trail of Gold and Silver*, 241–242, 234–235.

47. Adams, Dorsett, and Pulcipher, *Pioneer Western Bank*, vii.

48. "Eben Smith Brought First Stamp Mill into Colorado," 2; "Eben Smith, Pioneer and Mining Man Dies," 4; Application for accident insurance, February 3, 1897, Box 7A FF Smith, Eben Insurance (Accident Policy), Smith Papers.

Bibliography

Newspapers

Asheville (North Carolina) *Citizen*
Carbonate Chronicle (Leadville)
Central City (Colorado) *Register*
Daily Commonwealth (Denver)
Daily Miner's Register (Central City, Colorado)
Denver Post
Denver Republican
Denver Times
Erie (Pennsylvania) *Daily Times*
Evening Chronicle (Leadville)
Leadville Daily Herald
Los Angeles Herald
Los Angeles Times
Plaindealer (Ouray)

DOI: 10.5876/9781646421794.c010

Rocky Mountain Herald (Denver)
Rocky Mountain News (Denver)
Rocky Mountain Sun (Denver)
Sacramento Bee
Sacramento Daily Union
San Bernardino Sun
San Francisco Call
Sausalito (California) *News*
Steamboat (Springs) *Pilot*
Sunday Star (Washington, DC)

Manuscript Collections

First National Bank of Denver. Papers. Stephen H. Hart Library. History Colorado, Denver.

Georgetown and Breckenridge Wagon Road Company. Papers. Western History and Genealogy Department, Denver Public Library.

Jordan, Fletcher. "Jerome Chaffee and Eben Smith." Eben Smith Manuscript Series. Stephen H. Hart Library. History Colorado, Denver.

Moffat Estate Company. Papers. Western History and Genealogy Department. Denver Public Library.

Smith, Eben. Clipping File. Western History and Genealogy Department. Denver Public Library.

Smith, Eben. Papers. Western History and Genealogy Department. Denver Public Library.

Smith, Eben and Emily. Files. Gilpin Historical Society, Central City, CO.

Smith Family Papers. In possession of Michel Tritt. Private Collection.

Waite, Davis H. Papers. Colorado State Archives. Denver.

Watson and O'Kane Family Letters. Louis O'Kane Collection. Private Collection.

Wise, William. Scrapbook. Western History and Genealogy Department. Denver Public Library.

Other Sources

Abbott, Carl, Stephen J. Leonard, and David McComb. *Colorado: A History of the Centennial State*, 3rd ed. Boulder: University Press of Colorado, 1994.

Abbott, Carl, Stephen J. Leonard, and Thomas J. Noel. *Colorado: A History of the Centennial State*, 5th ed. Boulder: University Press of Colorado, 2013.

Adams, Eugene H., Lyle W. Dorsett, and Robert S. Pulcipher. *The Pioneer Western Bank: First of Denver, 1860–1980*. Denver: First Interstate Bank of Denver, 1984.

Arapahoe County Marriage Committee. *Marriages of Arapahoe County, Colorado, 1859–1901*. Denver: Colorado Genealogical Society, Inc., 1986.

Arkell, MacMillan, and Stewart, eds. *Aspen, Pitkin County, Colorado: Her Mines and Mineral Resources*. Aspen: *Aspen Daily Leader*, July 1, 1892.

Axford, H. William. *Gilpin County Gold: Peter McFarlane 1848–1929, Mining Entrepreneur in Central City, Colorado*. Chicago: Sage Books, 1976.

Bancroft's California Pioneer Registry and Index 1542–1848. Photocopy provided by Sacramento Public Library, Sacramento, CA.

Bradley, Christine, and Duane Smith. *The Once and Future Silver Queen of the Rockies: Georgetown, Colorado and the Fight for Survival into the Twentieth Century*. Louisville: University Press of Colorado, 2019.

Brands, H. W. *The Age of Gold: The California Gold Rush and the New American Dream*. New York: Doubleday, 2002.

Brown, Robert L. *Colorado Ghost Towns: Past and Present*. Caldwell, ID: Caxton, 2002.

Byers, William Newton. *Encyclopedia of Biography of Colorado, History of Colorado*. Chicago: Century Publishing and Engraving Co., 1901.

California, County Marriages, 1850–1952. www.familysearch.org, last accessed April 29, 2017.

California Death Index 1905–1939. www.familysearch.org, last accessed April 29, 2017.

California Death Index 1940–1997. www.familysearch.org, last accessed April 29, 2017.

Carnahan, Charles and Marie. Graves. Section S, lot 31. Riverside Cemetery. Asheville, NC.

Cone, Norris Hunter. Grave. Block 3. Fairmount Cemetery. Denver, CO.

Costello, Thomas. Passport Application 364517. January 8, 1924. familysearch.org, last accessed January 14, 2020.

Council Journal of the Legislative Assembly of the Territory of Colorado—Fifth Session. Central City, CO: David C. Collier, Printer, Miners' Register Office, 1866.

Council Journal of the Legislative Assembly of the Territory of Colorado—Sixth Session. Central City, CO: David C. Collier, Printer, Miners' Register Office, 1867.

Davis, John H. *The Guggenheims: An American Epic*. New York: William Morrow, 1978.

Denver City Directory, 1917, 1918, 1925, 1926, 1930. Denver, CO.

Eben Smith mausoleum. Block 3. Fairmount Cemetery. Denver, CO.

Edwards, Daniel W., and Roger W. Ward. *A Victorian Mansion in the Colorado Rockies: The Estemere Estate at Palmer Lake*. Privately published by the authors, 2012.

"Emergency Response to August 2015 Release from Gold King Mine." https://www.epa.gov/goldkingmine, last accessed February 17, 2020.

Fell, James E., Jr. *Ores to Metals: The Rocky Mountain Smelting Industry*. Boulder: University Press of Colorado, 2009.

Gladden, Sanford Charles. *Improvements in Boulder County*. Prepublication photocopy, Carnegie Branch, Boulder Public Library, 1987.

Granruth, Alan. *Mining Gold to Mining Wallets: Central City, Colorado 1859–1999*. Central City, CO: Gilpin Historical Society, 1999.

Gressley, Gene M. *West by East: The American West in the Gilded Age*. Charles Redd Monographs in Western History. Provo, UT: Brigham Young University Press, 1972.

Griswold, Don L., and Jean Harvey Griswold. *History of Leadville and Lake County Colorado: From Mountain Solitude to Metropolis*, vols. 1 and 2. Boulder: Colorado Historical Society and University Press of Colorado, 1996.

Hanna's Directory of California Landmarks. Photocopy provided by Sacramento Public Library, Sacramento, CA.

Hinton, Richard Josiah. *Rebel Invasion of Missouri and Kansas*. Chicago and Leavenworth, KS: Church and Goodman, and T. W. Marshall, 1865.

History of Clear Creek and Boulder Valleys, Colorado. Chicago: O. L. Baskin, 1880.

"History of the Mine and Smelter Supply Company." December 7, 1945. Western History and Genealogy Department, Denver Public Library. Denver, CO.

Inheritance Tax Record Book Three. Denver: City and County of Denver, July 28, 1910.

Iowa Mortality Schedule, 1850–1880. www.familysearch.org, last accessed April 17, 2019.

Jameson, Elizabeth. *All That Glitters: Class, Conflict, and Community in Cripple Creek*. Urbana: University of Illinois Press, 1998.

Jones, Helen Lukens. "The Greatest of Pipe Organs." *Scientific American* 90 (April 23, 1904): 328–329.

Lake County, Colorado, Marriage Records. Lake County Clerk and Recorders Office, Leadville, CO.

Leadville City Directories, 1883, 1885, 1890, 1891, 1892. Leadville, CO.

Lipsey, John J. *The Lives of James John Hagerman*. Denver: Golden Bell, 1968.

"The Long Strike at Cripple Creek: Complete History of the Miners' Struggle in the Great Gold Camp." *Great Divide* 11 (July 1894): 161–162.

Mann, Ralph. *After the Gold Rush: Society in Grass Valley and Nevada City, California, 1849–1870*. Stanford, CA: Stanford University Press, 1982.

Margaretta Smith Hickey et al v. the Estate of Emily Smith and Cora I. Costello. Case number 11628. Supreme Court of Colorado. Colorado State Archives. Denver.

Marshall, Thomas Maitland, ed. *Early Records of Gilpin County, Colorado, 1859–1861*. Denver: W. F. Robinson Printing Company, 1920.

McRoberts, Mary. *Boulder County, Colorado Censuses*. January 1992. Carnegie Branch. Boulder County Public Library. Boulder, CO.

Mehls, Steven F. *David H. Moffat, Jr.: Early Colorado Business Leader*. New York: Garland, 1989.

Mehls, Steven F. "Success on the Mining Frontier: David H. Moffat and Eben Smith, a Case Study." *Essays in Colorado History* 1 (1983): 92–105.

Mines and Mining Men of Colorado: Historical, Descriptive and Pictorial. Denver: John G. Canfield, 1893.

Moring, John. *Arthur Hill: Western Actor, Miner, and Law Officer*. Manhattan, KS: Sunflower University Press, 1994.

Mountain View Cemetery Records, 1857–1973. Oakland, CA. www.familysearch.org, last accessed February 19, 2020.

Noel, Tom, ed. *Glory That Was Gold*. Central City: Central City Opera House Association, 1992.

Pitts, John W., and Suzanne Pitts. *The Estemere Estate of Palmer Lake, 1887–1999*. Palmer Lake, CO: Palmer Lake Historical Society, 1999.

"Record of Appointment of Postmasters." 1832–September 30, 1971. Roll 14, Colorado, Colorado Territory, Adams–Kit Carson Counties. National Archives Microfilm Publications. Washington, DC.

Smith, Barbara Landweer. Telephone interview by the author, July 16, 2007.

Smith, Duane A. *The Birth of Colorado: A Civil War Perspective*. Norman: University of Oklahoma Press, 1989.

Smith, Duane A. "Chaffee: The Man behind the Name." *Colorado Central Magazine* 114 (August 2003): 19.

Smith, Duane A. *Henry M. Teller: Colorado's Grand Old Man*. Boulder: University Press of Colorado, 2002.

Smith, Duane A. *Horace Tabor: His Life and the Legend*. Boulder: Colorado Associated University Press, 1973.

Smith, Duane A. *Silver Saga: The Story of Caribou, Colorado*. Boulder: Pruett, 1974.

Smith, Duane A. *The Trail of Gold and Silver: Mining in Colorado, 1859–2009*. Boulder: University Press of Colorado, 2009.

Smith Family Plot. Plot 25, lot 5. Mountain View Cemetery. Oakland, CA.

Smith, William Eben. Death Certificate. File number 73–132001. October 16, 1973. Department of Public Health. State of California. Sacramento.

Stewart, John. *Thomas F. Walsh: Progressive Businessman and Colorado Mining Tycoon*. Boulder: University Press of Colorado, 2007.

Suggs, George G. *Colorado's War on Militant Unionism: James H. Peabody and the Western Federation of Miners*. Norman: University of Oklahoma Press, 1991.

Sze, Corinne P. "The Frank Leonard Smith House." *Bulletin of the Historic Santa Fe Foundation* 20 (November 1992): 1–13.

Tindall, George Brown, and David E. Shi. *America: A Narrative History*, 4th ed., vol. 2. New York: W. W. Norton, 1996.

Tutorow, Norman E. *Leland Stanford: Man of Many Careers*. Menlo Park, CA: Pacific Coast Publishers, 1971.

Ubbelohde, Carl, Maxine Benson, and Duane A. Smith. *A Colorado History*, 7th ed. Boulder: Pruett, 1995.

US Bureau of the Census. *Eighth Census of the United States, 1860*. Washington, DC: Government Printing Office.

US Bureau of the Census. *Fifteenth Census of the United States, 1930*. Washington, DC: Government Printing Office.

US Bureau of the Census. *Fourteenth Census of the United States, 1920*. Washington, DC: Government Printing Office.

US Bureau of the Census. *Seventh Census of the United States, 1850*. Washington, DC: Government Printing Office.

US Bureau of the Census. *Thirteenth Census of the United States, 1910*. Washington, DC: Government Printing Office.

US Bureau of the Census. *Twelfth Census of the United States, 1900*. Washington, DC: Government Printing Office.

Ward, Josiah M. "The Empire Builders of Colorado: David H. Moffat, the Banner Bearer." *Denver Post Magazine Section* (October 3, 1920): 2.

Waters, Frank. *Midas of the Rockies: The Story of Stratton and Cripple Creek*. New York: Covici, Friede, 1937.

White, B. S. *Representative Men of the West in Caricature*. Denver: *American Cartoonist Magazine*, 1904.

White, Richard. *"It's Your Misfortune and None of My Own": A New History of the American West*. Norman: University of Oklahoma Press, 1993.

Whitney, W. A., and R. I. Bonner. *History and Biographical Record of Lenawee County, Michigan*, vol. 1: *1879*. Adrian, MI: W. Stearns, 1879–1880.

Williams, Walter, ed. *A History of Northwest Missouri*, vol. 1. New York: Lewis, 1915.

Wright, B. Travis, and Kate Wright. *Images of America: Rollins Pass*. Charleston, SC: Arcadia, 2018.

Index